Proceedings of the

John Chappell

Natural Philosophy

Society

Volume 1

Copyright © 2015 by The John Chappell Natural Philosophy Society, Inc.

All rights reserved. No part of this publication may be reproduced, stored in a retrieval system, or transmitted, in any form, or by any means without prior permission in writing from the Natural Philosophy Alliance.

Printed in the United States of America.

ISBN 978-1-329-31311-8

PROCEEDINGS OF THE CNPS

Boca Raton, Florida 2015 (Vol. 1); © by the John Chappell Natural Philosophy Society
Published by the John Chappell Natural Philosophy Society, 1516 Crystalline Drive S.E., Caledonia, MI 49316, USA

The 1st Annual Conference of the CNPS, August 5-8, 2015 at Florida Atlantic University, Boca Raton, FL

This meeting was the first of our annual gatherings that took place at the Florida Atlantic University in Boca Raton Florida. It must be noted that the members of the John Chappell Natural Philosophy Society (CNPS) is made up of the 95% of the most active members of the NPA and have been meeting annual for over 20 years. We want to thank Cynthia Whitney and Greg Volk for their many years of dedication as editor of these proceedings, David de Hilster for putting on the conference and to David de Hilster for creating the 2015 Proceedings.

Proceedings of the CNPS, Volume 1

Acknowledgements

We want to thank Duncan Shaw, Lou Ellen LaFollette, Nick Percival, and Dr. Cynthia Whitney for making this conference possible.

More than that, Duncan Shaw Lou Ellen LaFollette, Greg Volk and Nick Percival were responsible for forming the new John Chappell Natural Philosophy Society (CNPS) in response to the immense dissatisfaction with the Natural Philosophy Alliance (NPA) directorate. The CNPS has been successful in attracting 95% of the active members from the NPA to the newly formed CNPS with bylaws giving members 100% control of the organization with the ability to remove anyone and everyone from the directorate if they see fit.

Without the people mentioned above, the CNPS, this conference, and these proceedings would not have been possible.

We also want to thank Bob, Pat, and Luanda de Hilster for their work in hosting the 1st CNPS conference in Boca Raton Florida.

- David de Hilster, Boca Raton, Florida

July 2015

An Explanation of Inertia Outside Relativity

Musa D. Abdullahi
12 Bujumbura Street, Wuse 2, Abuja, Nigeria
e-mail: musadab@outlook.com

An electric charge Q of mass m, in the form of a spherical shell of radius a, moving at time t with velocity **v** and acceleration $d\mathbf{v}/dt$, generates a dynamic electric field **X** proportional to the acceleration. The field **X** acts on the self-same charge Q to produce a reactive or inertial force $Q\mathbf{X} = -m(d\mathbf{v}/dt)$, equal and opposite to the accelerating force, in accordance with Newton's second law of motion, where m is a constant. This explains the cause of inertia as electrical and internal to a body, contrary to general relativity. An expression deduced for the mass m, in terms of charge Q and radius a, is compared with the electrostatic energy E of the charge to obtain the mass-energy equivalence equation, $E = \frac{1}{2} mc^2$, in contrast to the formula of special relativity, where c the speed of light in a vacuum.

Keyword: Acceleration, electric charge, electric field, force, inertia, mass, relativity, velocity.

1. Introduction

Newton's second law of motion defines force **F** in terms of acceleration $\frac{d\mathbf{v}}{dt}$ imparted to a particle of constant mass m, moving at time t with velocity **v**, and acceleration $d\mathbf{v}/dt$, as vector:

$$\mathbf{F} = m\frac{d\mathbf{v}}{dt} \quad (1)$$

In this case, inertia, the tendency of a body to resist being moved, that is accelerated or decelerated, becomes the reactive force or reverse effective force, equal and opposed to the accelerating force.

A moving charged particle carries along its own electrostatic field $\mathbf{E}_o$ and is associated with a magnetic field of intensity **H** perpendicular and proportional to its velocity relative to an observer. If the charge is accelerated linearly, it generates an electric field of intensity **X** proportional to the acceleration.

2. Magnetic field due a moving electric charge

A moving charged particle is associated with a magnetic field of intensity **H**, as shown in the Figure below. The magnetic field intensity **H** and magnetic flux intensity **B** are given by Biot and Savart law [1] of electromagnetism as vector (cross) product:

$$\mathbf{B} = \mu_o \mathbf{H} = \mu_o \varepsilon_o \mathbf{v} \times \mathbf{E}_o \quad (2)$$

where μ_o is the permeability, ε_o the permittivity of a vacuum and $\mathbf{E}_o$ is the electrostatic field of the moving charge. Equation (2) can be expressed as vector product:

$$\mathbf{B} = \mu_o \varepsilon_o \mathbf{v} \times \mathbf{E}_o = -\mu_o \varepsilon_o \mathbf{v} \times \nabla \varphi \quad (3)$$

where $\mathbf{E}_o = -\nabla \varphi$, φ is the potential at a point due to the charge, as given by Coulomb's law and ∇ denotes the 'gradient' of a scalar quantity. Transformation of equation (3) gives:

$$\mathbf{B} = -\mu_o \varepsilon_o \mathbf{v} \times \nabla \varphi = \mu_o \varepsilon_o \nabla \times \varphi \mathbf{v} \quad (4)$$

where $\nabla \times$ denotes the 'curl' of a vector and $\nabla \times \mathbf{v} = 0$.

3. Dynamic electric field due to acceleration

If the charge undergoes acceleration, a dynamic electric field of intensity **X** is generated, as given by Faraday's law of electromagnetic induction [2]:

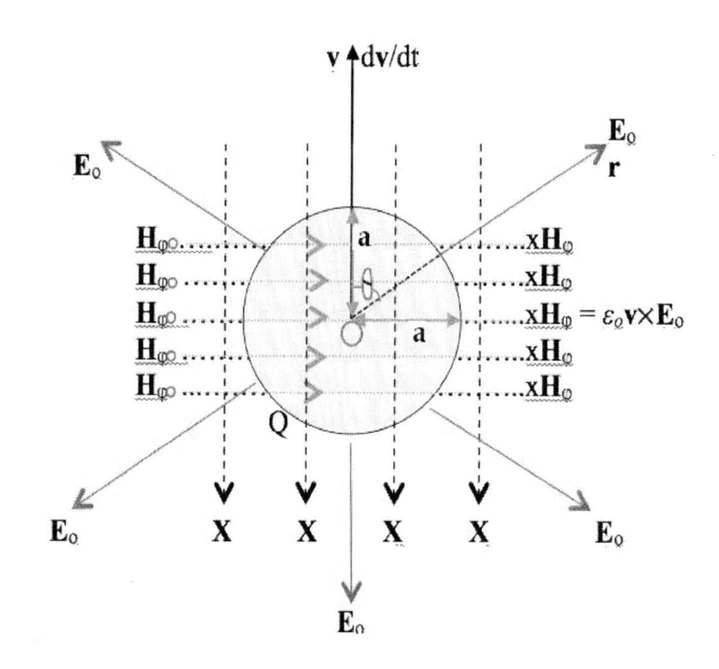

An electric charge Q and its radial electrostatic field E_o (red radial lines) moving at time t in a straight line with velocity **v** and acceleration $(d\mathbf{v}/dt)$ and generating a magnetic field **H** (out of the page on the left and into the page on the right) and-dynamic electric field of intensity **X** (dashed lines) in the opposite direction of acceleration.

$$\nabla \times \mathbf{X} = -\frac{d\mathbf{B}}{dt} \quad (5)$$

Since the potential φ is not a function of time t, Equation (4) and (5) give:

$$\nabla \times \mathbf{X} = -\frac{d\mathbf{B}}{dt} = -\mu_o \varepsilon_o \nabla \times \varphi \frac{d\mathbf{v}}{dt}$$

$$\mathbf{X} = -\mu_o \varepsilon_o \varphi \frac{d\mathbf{v}}{dt} = -\frac{\varphi}{c^2}\frac{d\mathbf{v}}{dt} \quad (6)$$

where $c = \sqrt{\dfrac{1}{\mu_o \varepsilon_o}}$ is the speed of light in a vacuum, a constant as determined by James Clerk Maxwell [3]. Equation (6) is the fundamental expression of this paper.

4. Inertial force on an accelerated charge

The fundamental assumption made here is that the dynamic dynamic field **X** acts on the self-same charge Q to produce the inertial force **F**, a reactive force which is equal and opposite to the accelerating force. Thus equation (6) and Newton's second and third laws of motion give the inertial force **F** as:

$$\mathbf{F} = \mathbf{X}Q = -\frac{\varphi Q}{c^2}\frac{d\mathbf{v}}{dt} = -\frac{UQ}{c^2}\frac{d\mathbf{v}}{dt} = -m\frac{d\mathbf{v}}{dt} \quad (7)$$

where mass of a particle m is considered as a constant independent of its speed v and U is the electrostatic potential at the point of location of the charge Q. At any position other than the point of location of the charge, the product UQ is zero. For a spherical charge of radius a, the intrinsic potential U is given by:

$$U = \frac{Q}{4\pi\varepsilon_o a} \quad (8)$$

The electrostatic energy of the charge, $w = UQ/2$, is the well-known formula for the electrostatic energy or intrinsic energy of a charge Q in its own electrostatic potential U, as given by:

$$w = \frac{Q^2}{8\pi\varepsilon_o a} \quad (9)$$

5. Mass-energy equivalence law

Equation (7), (8) and (9) give mass m and energy w as:

$$m = \frac{UQ}{c^2} = \frac{Q^2}{4\pi\varepsilon_o a c^2} = \frac{\mu_o Q^2}{4\pi a} \quad (10)$$

$$w = \frac{1}{2}mc^2 \quad (11)$$

For a mass m moving with speed v and kinetic energy $k = ½\, mv^2$, the sum E_n of the electrostatic and kinetic energies, is:

$$E_n = w + k = \frac{1}{2}m\left(c^2 + v^2\right) \quad (12)$$

6. Inertial force on a body

A body of mass M is composed of an equal number or equal quantities of $N/2$ positive (+Q) and $N/2$ negative (-Q) electric charges whose electrostatic fields balance out to zero at any point in space, in conformity with Gauss's law. The dynamic electric fields, as expressed in equation (6), act on the respective charges of total mass $M = Nm$ so that equation (7) becomes:

$$\mathbf{X}NQ = -\frac{UNQ}{c^2}\frac{d\mathbf{v}}{dt} = -Nm\frac{d\mathbf{v}}{dt} \quad (13)$$

Equation (11), for a body of mass M, is obtained as:

$$W = \frac{1}{2}Mc^2 \quad (14)$$

where W is the total electrostatic energy of the charges constituting the body of mass M. For a body of mass M moving with speed v and kinetic energy $K = ½\, mv^2$, the total energy E, is:

$$E = W + K = \frac{1}{2}M\left(c^2 + v^2\right) \quad (15)$$

7. Discussions

Newton's second law of motion, Coulomb's law of electrostatic force and basic electrostatic, electromagnetic and electrodynamic principles are employed to explain the cause of inertia (equation 7) and to derive equations (12) and (15), without recourse to the theories of relativity or any other principle. The assumption made is that the dynamic electric field **X** produced by an accelerated charge acts on the self-same charge Q to produce the inertial force **X**Q equal and opposite to the accelerating force. This interaction takes place in so far as an electric field always exerts a force on an electric charge even where the field is generated by the same charge.

The assumption here is backed by Newton's second and third laws of motion and Faraday's law of electromagnetic induction. For a body composed of equal numbers of positive and negative electric charges, the individual dynamic electric fields act on their respective charges, at their individual locations, to produce the inertial force. The dynamic electric fields, like the electrostatic fields, of a neutral body, cancel out externally.

There is a similarity in the explanations of inertia and gravity [4]. Both are the results of electric fields - a dynamic electric field and an "electro-gravity" field - acting on a charge. Inertia and gravitation are internal properties of a body by virtue of its composition with equal numbers or equal amounts of positive and negative charges.

8. Conclusion

This paper is based on the use of intrinsic energy U and mass m of a particle of charge Q, in the form of a spherical shell of radius a, to obtain equation (10). As a consequence, since electric charge is a constant, mass of the charge should also be a constant independent of speed, contrary to special relativity.

Another consequence of the explanation given here makes inertia electrical in nature and a property residing in a body. The derivation of a mass-energy equivalence law ($E = ½\, mc^2$) is quite straightforward but differs from the relativistic formula ($E = mc^2$) by a factor of one half. Whatever the case may be, E being the electrostatic energy of the electric charges constituting a body, makes things easier and brings a new insight in electrodynamics.

References

[1] I.S. Grant & W.R. Phillips; *Electromagnetism*, John Wiley & Sons, New York (2000), p. 137-8.

[2] D.J. Griffith; *Introduction to Electrodynamics*, Prentice-Hall, Englewood Cliff, New Jersey (1981), pp. 257 – 260.

[3] J.C. Maxwell; *A Treatise on Electricity and Magnetism*, 3rd Ed. (1892), Part IV, Chap.2.

[4] M.D. Abdullahi (2004); "Explanation of gravity outside relativity, (unpublished).

[5] http://gsjournal.net/Science-Journals/%7B$cat_name%7D/View/5564

Terminal Speed of an Electron Accelerated by an Electric Field

Musa D. Abdullahi
12 Bujumbura Street, Wuse 2, Abuja, Nigeria
e-mail: musadab@outlook.com

An electron of mass m and charge $-e$ moving at time t in a straight line with speed v and acceleration dv/dt in an electric field of magnitude E, comes under an electrostatic or impressed force $-eE$. It is proposed that the electron also encounters a radiation reaction force, a kind of frictional force $-eEv/c$ opposing motion. The accelerating force becomes $F = -eE(1 - v/c) = -m(dv/dt)$, in accordance with Newton's second law of motion, where c is the speed of light in a vacuum. Thus the accelerating force F decreases on speed v in the field, contrary to Coulomb's law. The electron emits radiation at power eEv^2/c as it reaches a terminal speed equal to that of light c, with zero accelerating force when the impressed force becomes equal and opposite to the radiation reaction force.

Keywords: Acceleration, force, electric charge, field, mass, speed, relativity

1. Introduction

In a remarkable experiment in 1964, William Bertozzi [1] of the Massachusetts Institute of Technology demonstrated the existence of a universal limiting speed, equal to the speed of light c in a vacuum. The experiment showed that electrons accelerated through energies over 15 MeV, attain for all practical purposes, the speed of light $c \approx 2.998 \times 10^8$ meters per second. Linear particle accelerators of energies over 100 MeV have been operated with electrons moving practically at the speed of light.

The theory of special relativity [2] explains the existence of a limiting speed equal to the speed of light by positing that the mass of a moving particle, such as an electron, increases with its speed, becoming infinitely large at the speed of light. Since an infinite mass cannot be accelerated any faster by a finite force, the speed of light c should reasonably be the ultimate limit.

This paper shows that the accelerating force exerted by an electric field on a moving electron decreases with its speed, reducing to zero at the speed of light. Infinite mass or zero force at the speed of light leads to zero acceleration and constant speed c as a limit, in accordance with Newton's laws of motion. It is unfortunate that special relativity chose *mass* as the variable rather than *force*. The ultimate speed without infinite mass is a more comfortable and more realistic proposition [3].

Radiation reaction force due to an electron of charge $-e$ moving with acceleration in a straight line at velocity $\mathbf{v}$ in the opposite direction of an electric field of intensity $\mathbf{E}$ is put as $-eE\mathbf{v}/c = e\mathbf{E}v/c$. In contrast to Coulomb's law, the accelerating force is the sum of the impressed force (the electrostatic force) $-e\mathbf{E}$ and the radiation reaction force $e\mathbf{E}v/c$. At the speed of light, the accelerating force, $-e\mathbf{E} + e\mathbf{E}v/c$ on the electron, reduces to zero and the particle moves with zero acceleration and constant speed c.

2. Acceleration in rectilinear motion

An electron of mass m and charge $-e$ moving at time t, in a straight line, with velocity $\mathbf{v}$ and acceleration $(d\mathbf{v}/dt)$ in the opposite direction of an electric field of intensity E, encounters a radiation reaction force put as $-eE\mathbf{v}/c = e\mathbf{E}v/c$. The accelerating force F on the electron, equal to the sum of the impressed force $-e\mathbf{E}$ and the radiation reaction force $e\mathbf{E}v/c$, that is $-e\mathbf{E}(1 - v/c)$, is given by Newton's second law of motion, as vector equation:

$$\mathbf{F} = -e\mathbf{E}\left(1 - \frac{v}{c}\right) = m\frac{d\mathbf{v}}{dt} \quad (1)$$

Since velocity $\mathbf{v}$, of magnitude v, is in the opposite direction of the electric field $\mathbf{E}$, the scalar equation is:

$$F = -eE\left(1 - \frac{v}{c}\right) = -m\frac{dv}{dt} \quad (2)$$

where c is the speed of light in a vacuum and m is the mass of the particle, which is considered to be independent of speed v of the particle. In other words, mass m of a moving particle remains equal to the rest mass m_o.

For acceleration in a uniform field (E constant), where $eE/m = a$ is a constant, the solution of equation (2), for an electron accelerated from an initial speed u, is:

$$v = c - (c - u)\exp\left(\frac{-at}{c}\right) \quad (3)$$

For an electron accelerated from an initial speed $v = 0$, the solution of equation (2) is:

$$\frac{v}{c} = 1 - \exp\left(-\frac{at}{c}\right) \quad (4)$$

In equations (3) and (4), the speed of light c is the limit to which a charged particle, such as an electron, can be accelerated by an electric field. A graph of v/c against at/c is shown as curve *C1* in Figure 1 below for equation (4).

3. Deceleration in rectilinear motion

For a decelerated electron the differential equation of motion (replacing v by $-v$ in equation 2) becomes:

$$F = -eE\left(1+\frac{v}{c}\right) = m\frac{dv}{dt} \tag{5}$$

The solution of equation (5), for a charged particle decelerated, by a uniform electric field, from initial speed u, is:

$$v = (c+u)\exp\left(-\frac{at}{c}\right) - c \tag{6}$$

For deceleration from the speed of light, $u = c$, equation (6) becomes:

$$\frac{v}{c} = 2\exp\left(-\frac{at}{c}\right) - 1 \tag{7}$$

In equations (6) and (7), the particle is decelerated to a stop and then accelerated in the opposite direction to reach a terminal speed equal to $-c$, as shown in curve *C2* of the graphs in Figure 1. This result is not obtainable from the point of view of the theory of special relativity. In special relativity a particle moving at the speed of light will continue to move at that speed, losing potential energy without gaining kinetic energy.

4. Speed-time Equations and of motion

Figure 1 is a graph of speed versus time for an electron accelerated from zero speed or decelerated from the speed of light c.

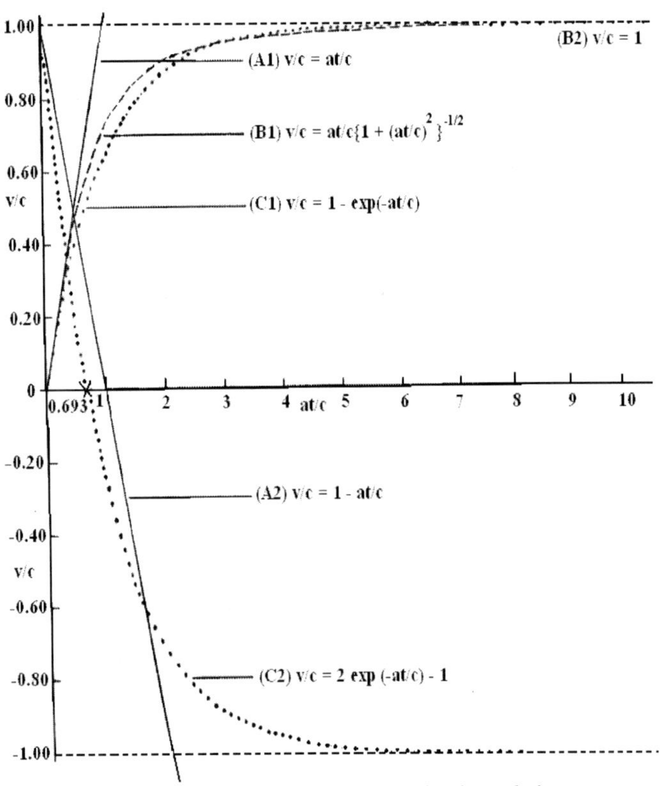

Fig. 1: Speed versus time for an accelerated or decelerated electron

Figure 1 is a graph of v/c *(speed in units of c)* against at/c *(time in units of c/a)* for an electron of charge $-e$ and mass m equal to the rest mass m_0 accelerated from zero initial speed or decelerated from the speed of light c, by a uniform electric field of magnitude E, where $a = eE/m$; the lines(A1) and (A2) according to classical electrodynamics, the dashed curve (B1) and line (B2) according to relativistic electrodynamics and the dotted curves (C1) and (C2) according to equations 4 and 7.

In classical electrodynamics, an electron of charge $-e$ and mass m, moving at the speed of light c, on entering a uniform retarding field of magnitude E, should be stopped in time $t = mc/eE$, energy radiation notwithstanding. In relativistic electrodynamics an electron moving at the speed of light c should be unstoppable by any force. In the alternative electrodynamics, an electron moving at the speed of light c is easily decelerated to a stop, by a uniform retarding field E, in time $t = 0.693mc/eE$, with radiation of energy.

5. Conclusions

Equations (2) and (5) are simply extensions of Coulomb's law of electrostatic force taking into consideration the speed of a charged particle in an electric field. Equations (4) and (7) give the speed of light c as the ultimate speed with mass of a moving particle remaining constant as the rest mass. It is energy radiation that makes all the difference between classical, relativistic and the alternative electrodynamics advanced here. Making Coulomb's law independent of velocity of a charged particle in an electric field is a long-standing mistake.

The speed of light c, being an ultimate limit to which a charged particle can be accelerated by an electric field, has nothing to do with the mass of the particle. It is a property of the electric field and it is radiation reaction force, a kind of frictional force proportional to speed, which limits the speed of an accelerated charged particle to that of light c.

References

[1] W. Bertozzi; "Speed and Kinetic Energy of Relativistic Electrons", *Am. J. Phys.*, 32 (1964), 551–555. Also online at: http://spiff.rit.edu/classes/phys314/lectures/relmom/bertozzi.htm

[2] A. Einstein; "On the Electrodynamics of Moving Bodies", *Ann. Phys.*, 17 (1905), 891.

[3] http://www.musada.net/Papers/Paper1.pdf

Force in Boscovich's Unified Field Theory

Roger J Anderton
England, UK
e-mail: R.J.Anderton@btinternet.com

After the Copernican revolution modern physics accepted atomism, i.e. that matter was composed of point-particles of Boscovich's theory. Rarely is it mentioned by physicists where this concept came from, and even rarer is it mentioned the unified theory based on it. Einstein's highly influential impact on 20th Century physics seems to have diverted attention from the fact that there was a unified field theory before his influence. Einstein faced the problem of how could electromagnetism and gravity be unified in his space-time. In the context of Boscovich's theory unification is much simpler.

1. Introduction

The Physics Establishment has sadly not paid much attention to its history, and as it struggles to try to unify general relativity with quantum mechanics, usually forgets about Boscovich's unified theory.

2. 18th Century Unified Field Theory

Lancelot Law Whyte [1] points out that Roger Boscovich published in 1758 "the first general mathematical theory of atomism, based on the ideas of Newton and Leibniz but transforming them so as to provide a program for atomic physics."

Since atomic physics is now so important, we should pay attention to Boscovich's theory as to where that physics came from. In this theory we deal with "centers of force" as the fundamental particles. This concept arises naturally from Newton's universal law of gravitation ($F= GMm/r^2$) where we consider gravity as acting between the centers of mass M and m, through distance r. (Care must of course be taken with density for these point-particles where mass of an object acts as if it were from the center of the object.)

Boscovich goes further than Newton to give a universal force, where the point-centers "obey an oscillatory (alternatively repulsive and attractive) law of interaction, dependent on the distance between each pair [of point-centers]." Whyte points out various problems with what he conceives in Boscovich's theory. However, I content that most of those problems are due to Einstein who has had a massive influence of modern theoretical physics and introduced many errors making things look different to how Boscovich conceived of things. Modern atomic physics remains built upon Newtonian physics from the insights of Boscovich, despite Einstein's influence.

Of the many insights and changes that Einstein made that leads to numerous problems, is the one that is recorded in various texts as: "One of Albert Einstein's greatest insights was realizing that time is relative. It speeds up or slows down depending on how fast one thing is moving relative to something else." [2]

As Nick Percival [3] points out: "Einstein redefined "time" and he did it with virtually no discussion of why "time" needed redefining or even that he was making a major change in current thinking on the metaphysics of "time". And Einstein redefined "time" in terms of the speed of light no less!!! Even more shocking, the physics world accepted Special Relativity, albeit slowly, with virtually no discussion of this major redefinition of "time"! It should have been clear to all that the 1905 paper's implicit claim that "time" itself changes as a function of relative velocity leads to immediate and obvious contradictions as soon as one considers multiple observers."

Time dilation in the context of special relativity is derived from the following right angled Pythagorean triangle [4] as hypotenuse = ct, vertical = L and horizontal = vt , where c= speed of light, v = velocity that frame of reference is moving, t = time observed. Issues to be considered is that it is formed in idealization of an inertial frame, empty space etc.

From this we have: vertical $L = \sqrt{(c^2 - v^2)} * t$

Which is then made equal to ct', and after math manipulation gives the time dilation equation.

If it were Newtonian Physics then it would be: $t = t'$ and $c' = \sqrt{(c^2 - v^2)}$.

So we have in general the equation: $c' * t' = \sqrt{(c^2 - v^2)} * t$.

Einsteinians take: $c' = c$ and gets t different to t'.

But should really do things by Newtonian physics where $t' = t$, then $c' = \sqrt{(c^2 - v^2)}$.

There is no justification as to why Einsteinians make the change. It means that Einsteinians are just imposing on experiments that lightspeed should be interpreted as constant so that there are effects like time dilation; whereas from the maths they use we can see it can still be interpreted in the old Newtonian way. And in the context of carrying on with Newtonian physics we can carry on with its way of dealing with forces, there being no need for the change to space-time curvature et al., which comes from further mistakes that the Einsteinians make.

We then have the electromagnetic force (in vector notation) as [5]: $\underline{F}_{Lorentz} = q\,\underline{E} + q\,\underline{v} \times \underline{B}$ ....(1)

This is the Lorentz force exerted on a charged particle q moving with velocity $\underline{v}$ through an electric field $\underline{E}$ and magnetic field $\underline{B}$.

Now that we are dealing with forces as per Newtonian physics, we need merely to include gravity by adding the equation [6]: $GMm/r^2 = m\,a$ ....(2)

G = Gravitational constant, M = mass of first object, m = mass of second object, r = distance between centers of objects, a = magnitude of acceleration for object with mass m.

Thus for the total force of electromagnetism and gravity (treating everything as magnitudes) we have:

$F_{total} = q\,E + q\,v\,B + m\,a$ ....(3)

This in Boscovich's theory obeys his oscillatory law of universal force as per his force curve, [7] the details of which seem approximately along the lines of a radial inverse cube repulsive force and radial inverse square attractive force. In the context of Boscovich's theory we are dealing with point-particles (or centers of force), an approach that dates back to the Pythagoreans. As Morris Kline points out [8]: "The early Pythagorean doctrine is puzzling because to us numbers are abstract ideas, and things are physical objects or substance. But we have made an abstraction of number which the early Pythagoreans did not make. To them, numbers were points or particles." i.e. Boscovich's theory is unified theory based on these Pythagorean point-particles. This was a natural progression of the Copernican revolution, because Copernicus was deemed by the Church as working due to the ancient Pythagorean doctrine. [9]

3. Conclusion

It was as simple as this to form the unified force law, something that Einstein apparently had so many difficulties with when he did his search for unified field theory. From his perspective he was dealing with electromagnetic force in the flat space-time of special relativity, and then he dealt with the Newtonian gravitational force as space-time curvature in general relativity, and by so doing make more mistakes. The two different ways of representing force was then hence confusing, and he tried to reconcile them. Now going by there was no need to represent time as different (i.e. can carry on using t =t', instead of having them different), we can carry on treating force in the old pre-Einstein way. Thus we need to restart the Newtonian research program.

References

[1] Whyte, Lancelot Law, 1961 Essay On Atomism From Democritus To 1960, Nelson, p. 54-55

[2] Nova Science, Einstein's Big idea, http://www.pbs.org/wgbh/nova/einstein/hotsciencetwin/

[3] Percival, Nick, Jun 1, 2015, The Physics of the Twin Paradox, http://www.naturalphilosophy.org/site/nickpercival/2015/06/01/the-physics-of-the-twin-paradox/

[4] For further details as to how this is formed see such places as: http://www.drphysics.com/syllabus/time/time.html

[5] Encyclopaedia Britannica, Lorentz force, http://www.britannica.com/EBchecked/topic/348139/Lorentz-force

[6] https://www.wyzant.com/resources/lessons/science/physics/gravitation

[7] Stoiljkovich, Dragoslav, trans Roger Anderton, Roger Boscovich: The founder of modern science, ISBN 978-86-7861-043-1

[8] Kline, Morris, Mathematics: The Loss of Certainty, ISBN-13 978-0195030853,p.12

[9] Adriano (di St. Thecla), Olga Dror, Opusculum de Sectis Apud Sinenses Et Tunkinenses, ISBN 0-87727-732-X, 2002, p.57

Introducing Integral Geometry: Are Notational Flaws Responsible For the Inability to Combine General Relativity and Quantum Mechanics?

Jeffrey P. Baugher
e-mail: jpbaughernps@gmail.com

Parallel line segments are the basic graphical foundation for geometrical field theories such as General Relativity. Although the concept of parallel and curved lines have been well researched for over a century as a description of gravity, certain controversial issues have persisted, namely point singularities (Black Holes) and the physical interpretation of a scalar multiple of the metric Λ, commonly known as a Cosmological Constant. We introduce a graphical and notational analysis system which we will refer to as Integral Geometry. Through variational analysis of perpendicular line segments we derive equations that ultimately result from the changes in the area bounded by them. Based upon changing area bounded by relative and absolute line segments we attempt to prove the following hypothesis: General Relativity cannot be derived from Integral Geometry. We submit that examination of the notational differences between GR and IG in order to accept the hypothesis could lead to evidence that the inability to merge General Relativity and Quantum Physics may be due to notational and conceptual flaws concerning area inherent in the equations describing them.

1. Introduction

In this work, we introduce the concept of Integral Geometry (IG). This concept is an examination of relative and absolute areas, the resulting equations from their summations and differences and finally, physical modeling of differential absolute and relative areas based on perfect fluids and spatial and temporal probabilities. We have found that absolute areas seem to be suitable for building absolute coordinate systems for which we can track particles and that relative areas are suitable for tracking relative waves through a perfect fluid.

In order to facilitate appreciation of some of the possibilities of IG, it is necessary to understand the similarities and differences with current physical laws and equations. So as to keep this as simple and compact as possible, we focus on understanding the similarities of the Cosmological Constant (CC or Λ) problem and that of constant relative area within Integral Geometry.

The CC from General Relativity (GR) has several different names: a Cosmological Constant, a scalar multiple of the metric, an Einstein manifold and a postulated energy density of the vacuum. The "problem" [1] stems from the fact that although this constant seems to be present both in quantum mechanics (QM) and GR, the estimated value within QM is over 100 orders of magnitude different than what would seem to work within GR from examination of the actual empirical evidence.

This problem is also illustrated by a central equation within metric field theories where the effect that a metric $g_{\mu\nu}$ has on space-time $dx_\mu dx_\nu$ is described by [2]

$$ds^2 = g_{\mu\nu} dx_\mu dx_\nu.$$

The view within these field theories is that a metric changing from point to point can describe a non-Euclidean "curvature" of space-time. The Einstein field equation

$$R_{\mu\nu} - \frac{1}{2} g_{\mu\nu} R = \frac{8\pi G}{c^4} T_{\mu\nu}$$

would then be a description of how energy ρ and momentum p (within $T_{\mu\nu}$) effect $g_{\mu\nu}$. If there is no momentum or energy present, then the equation

$$R_{\mu\nu} = 0$$

would describe zero curvature and a summation of the unchanged components of $g_{\mu\nu}$ would each have a magnitude of $|1|$ and be written as

$$R = 4.$$

There has been, however, the conceptual discrepancy of the CC which would have summed components of

$$R = 4\Lambda.$$

This is normally just added into the Einstein field equation as

$$R_{\mu\nu} - \frac{1}{2} g_{\mu\nu} R + \Lambda g_{\mu\nu} = \frac{8\pi G}{c^4} T_{\mu\nu}$$

but we stress that there is no known way to calculate a theoretical value for Λ that would match any proposed physical explanation. As an example, what would

$$ds^2 = \Lambda g_{\mu v} dx_{\mu} dx_{v}$$

mean?

For comparison purposes, we derive here a conceptual framework within IG which we have named Line Segment Space (LSS). During our preliminary research we have developed the equation in LSS of

$$\frac{S_v}{dS_v} dS_v dS_h = S_v dS_h.$$

If we investigate defining the first fraction as

$$\frac{S\mu}{dS\mu} \equiv \Lambda g_{\mu v}, \mu = v, v = h$$

we then examine why we can find LSS solutions for which

$$|S_{00}| + |S_{11}| + |S_{22}| + |S_{33}| = 0$$

and

$$\frac{\left(dS_{\mu}\right)^2}{(dS_v)^2} = 0$$

But *no* metric solutions for

$$R = 0$$

And

$$R_{\mu v} = 0.$$

We propose that this investigation may unveil serious questions of whether not only metric field theory is conceptually valid but Euclidean and non-Euclidean geometry itself. For example, if $S_h = f(S_v)$, can we use this geometric model also as a physical model where relative changes in x and t are functions of relative changes in the density ρ and pressure p of the vacuum when treated as a perfect fluid?

2. Derivation

We keep the derivation of this hypothesis simple and non-robust so as to require consideration of the widest range of interpretations to verify the following hypothesis: General Relativity cannot be derived from Integral Geometry.

IG contains three separate conceptual regions: Graphical, Notational and Physical. We demonstrate on how to examine a single concept as it evolves through all three regions.

A. Graphical-Visual examination of proofs.

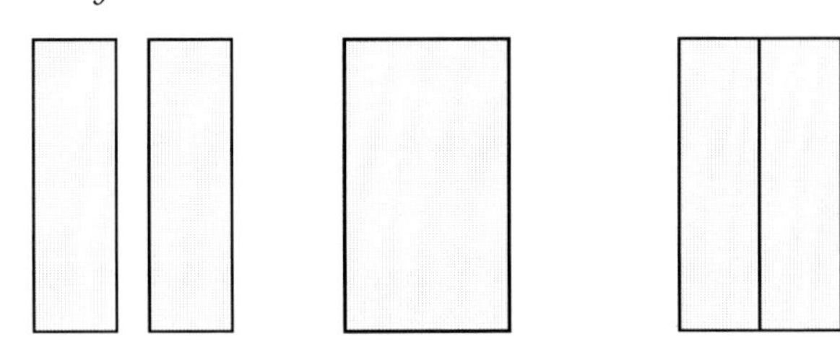

Figure 1. Summation of two areas on left equal single area in middle.

B. Notational-Notational descriptions of graphical proofs.

$$S_v^1 S_h^1 + S_v^2 S_h^2 = S_{total}$$

$$S_{total} - S_v^1 S_h^1 + S_v^2 S_h^2 = 0$$

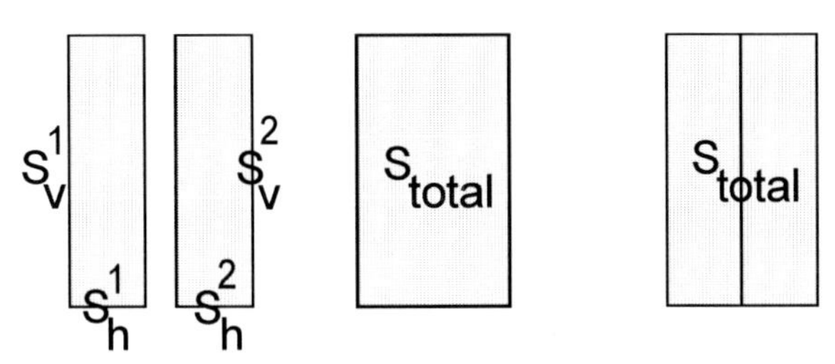

In IG, there are two basic notational forms, line segment notation and point notation. $\boldsymbol{S}$ is a single descriptor of a line segment. $\boldsymbol{A_1} - \boldsymbol{A_2}$ is a point descriptor of a line segment, such that $\boldsymbol{S} = \boldsymbol{A_1} - \boldsymbol{A_2}$. This distinction will become important in the later discussion of the notation within various historical metric field theories.

C. Physical

Let us consider the issue of equating notational descriptions to physical models. Current physics cannot find a workable relationship between the Poisson equation describing energy and the Cosmological Constant as the energy density of the vacuum. Through attempts at falsifying our hypothesis, we propose to examine whether this difficulty is due to a notational system within GR that inherently accounts only for parallel and curved lines instead of a more appropriate view of changing relative area.

$$\nabla^2 \phi = \rho$$

$$\Lambda g_{\mu v} = \rho_{vac}?$$

3. Geometric Summation of Area

A. Quantities of Area Can Be Summed

$$\sum S_v^m S_h^n = S_v^1 S_h^1 + S_v^2 S_h^2 + S_v^m S_h^n$$

B. Defining Integration as Summation of Infinitesimal Areas

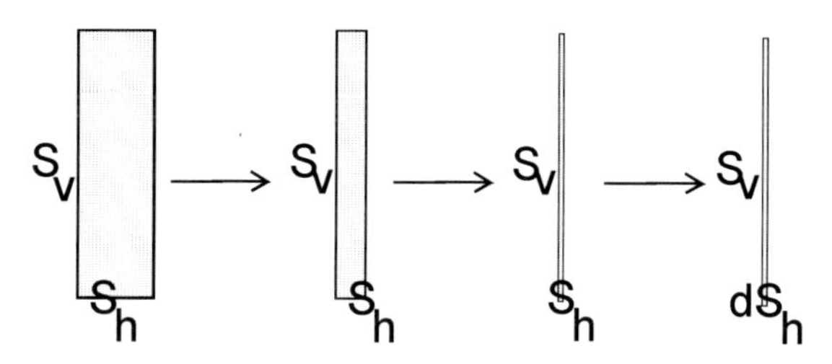

Figure 2. Rectangle of Area Becomes an Infinitesimal Slice of Area

Summation of blocks of area becomes integration of infinitesimal slices. We specifically point out that *no functional relationship between the line segments is required for either summation or integration of area.* We call the horizontal line segment of zero width dS_h a *point derivative.*

$$\sum S_v^m S_h^n \rightarrow \int S_v^m dS_h^n$$

$$S_v \neq f(S_h)$$

$$S_h \neq f(S_v)$$

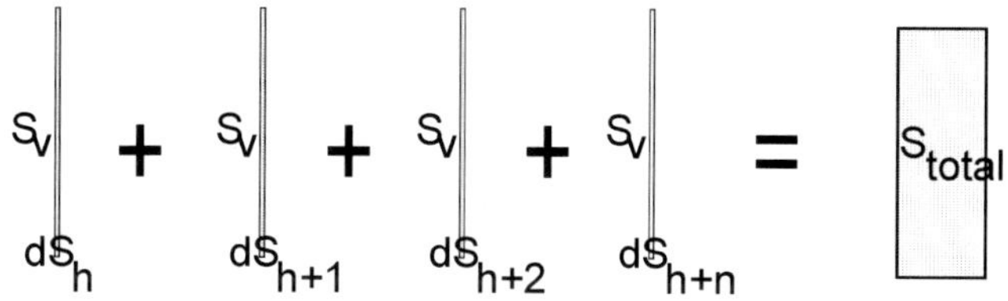

Figure 3. Integration of Infinitesimal Slices of Area

4. Differences of Infinitesimal Slices of Area Are Notationally Definable

Differences of area are definable therefore differences of infinitesimal slices of area are definable.

$$S_v^1 S_h^1 - S_v^2 S_h^2 \neq 0$$

$$S_v^1 S_h^1 - S_v^2 S_h^2 = (S_v^1 - S_v^2) S_h \neq 0$$

$$(S_v^1 - S_v^2) \neq 0$$

5. Infinitesimal Differences of Infinitesimal Slices of Area are Definable

We can define an infinitesimal limit of when the differences between the lengths of the vertical line segments go to zero. We call the infinitesimal comparison of the vertical line segments dS_v a *line derivative*. It is important to note that neither line segment is required to have, but could have a magnitude of 0.

$$|S_v^1| \neq 0$$

$$|S_v^2| \neq 0$$

$$\lim_{\Delta \; between \; S_v^1 \; and \; S_v^2 \rightarrow 0} (S_v^1 - S_v^2) dS_h = dS_v dS_h$$

The special case

$$dS_v dS_h = 0$$

exists when

$$|S_v^1| = |S_v^2|.$$

6. The Relative Rate of Change of Area Can be Defined Through Normalization

We can consider values of ratios of line derivatives and point derivatives or consider solutions when we consider these ratios to be constant.

$$\frac{dS_v dS_h}{dS_h dS_h} = \frac{dS_v}{dS_h} \neq 0$$

$$\frac{dS_v dS_v}{dS_h dS_h} = \frac{dS_v^2}{d^2 S_h} \neq 0$$

7. Introduction of Line Segment Space

We define Line Segment Space (LSS) as a row of an infinite number of points. We can think of each point having a vertical line segment S_v.

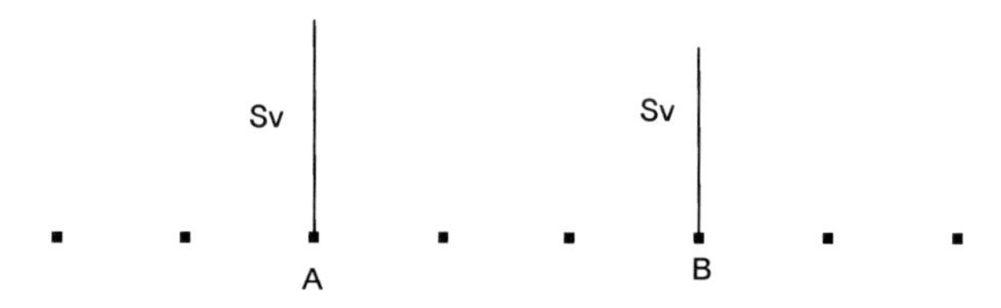

Figure 4. Infinite number of points in a row each with its own S_v.

For S_h however, there are two paradigms, absoluteness and relativity.

A. Absolute Line Segment in LSS

Any magnitude or section of S_h is made up of multiple points on the row of points, not necessarily of the same magnitudes. The area bounded by these line segments cannot conceptually overlap.

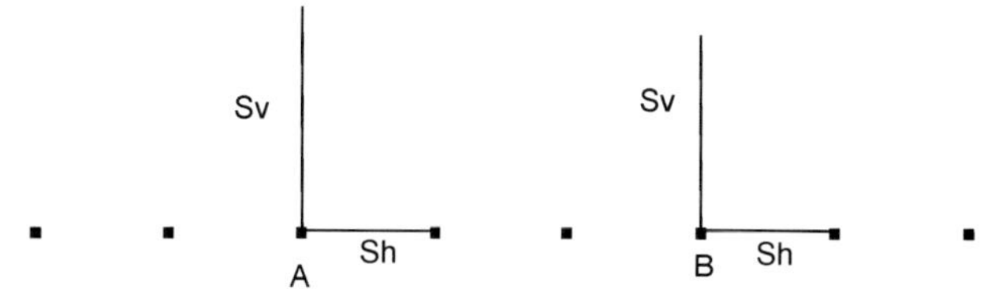

Figure 5. Any S_h is made up of points in the row.

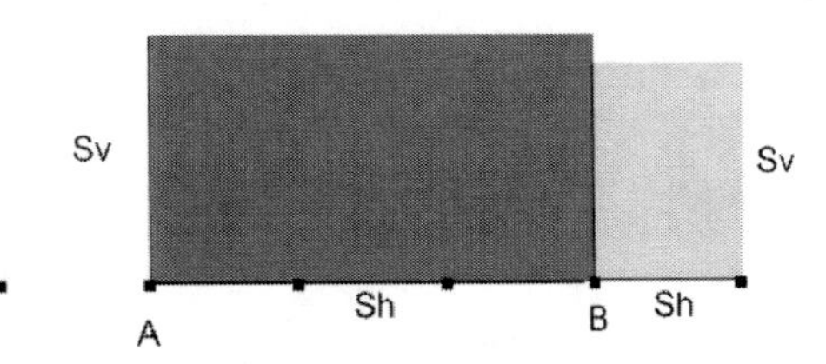

Figure 6. Areas cannot conceptually overlap.

B. Relative Line Segment in LSS

Each point in the row of points has a separate magnitude or section of S_h, not necessarily of the same magnitudes.

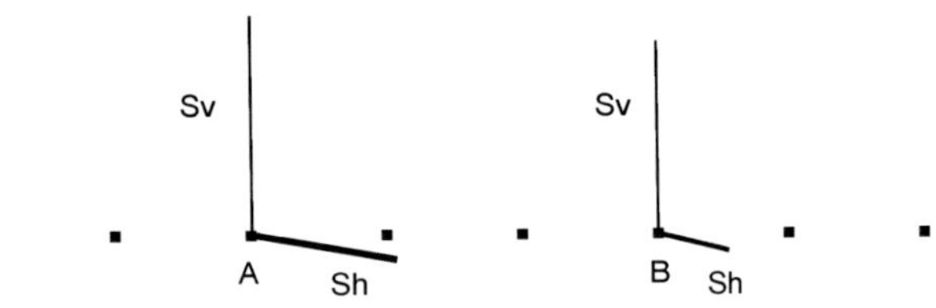

Figure 7. Every point contains a line segment S_h.

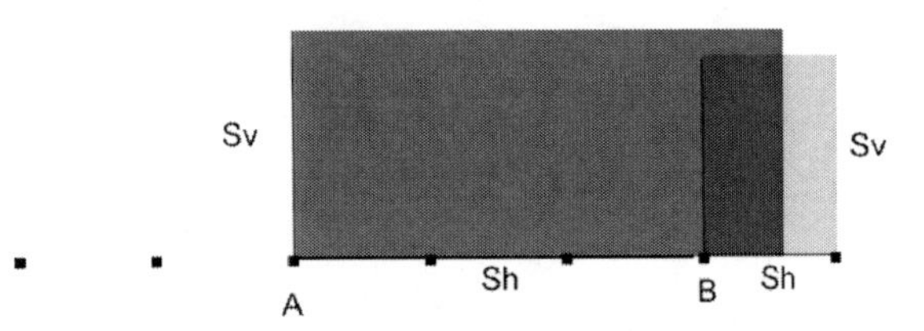

Figure 8. Areas can conceptually overlap.

C. Functional Relationships Between S_h and S_v and Physical Interpretations

We consider these questions of the two simplest functional relationships between S_h and S_v:

For absolute line segments, can the Poisson equation

$$\nabla^2\phi = \rho$$

be derived from

$$\frac{dS_v dS_v}{dS_h dS_h} = \frac{dS_v^2}{d^2S_h} = b$$

and

$$\phi \equiv S_v$$

via

$$S_v = f(S_h)?$$

For relative line segments, can

$$ds^2 = g_{\mu\nu}dx_\mu dx_\nu$$

be derived from

$$\frac{S_v}{dS_v}dS_v dS_h = S_v dS_h$$

and

$$\frac{S_\mu}{dS_\mu} \equiv \Lambda g_{\mu\nu}$$

via

$$S_\mu = f(S_\nu)?$$

8. Motivation

a. Is an Einstein Manifold the Same as Constant Non-zero Relative Area in LSS?

i. In LSS using relative S_h, for any S_v that is the same at point A and point B in the row, $\left(\frac{S_v}{dS_v}dS_v dS_h\right)_A - \left(\frac{S_v}{dS_v}dS_v dS_h\right)_B = 0.$

ii. We currently can find no way to falsify this equation from being derived through $\frac{S_\mu}{dS_\mu} \equiv \Lambda g_{\mu\nu}$ into either $Rg_{\mu\nu} = 0$ or $R\Lambda g_{\mu\nu} = 0$ with the exception that solutions for $|g_{\mu\nu}| = 0$ with $Rg_{\mu\nu} = 0$ do not conceptually exist in metric field theory.

iii. Is allowing the diagonal arguments to be $|g_{\mu\nu}| = (1,1,1,1)$ a *specific* understanding of relativity rather than a *general* one considering that any constant for Λ satisfies the equation $R\Lambda g_{\mu\nu} = 0$?

iv. As examples, constants $|\Lambda g_{\mu\nu}| = (5,5,5,5)$ or even $|\Lambda g_{\mu\nu}| = \left(\frac{198765}{35467}, \frac{198765}{35467}, \frac{198765}{35467}, \frac{198765}{35467}\right)$ at all points satisfies the equation. Why is the notation accepted if the solutions are not unique?

b. Is a Point Singularity in GR Conceptually Equivalent to Running Out of Relative Area in LSS?

c. Is a second set of four time-space components within $ds^2 = g_{\mu\nu}dx_\mu dx_\nu$ (dx_μ first set, dx_ν second set) a demonstration of incorrect notation for a vertical line segment that has its own point derivative included?

d. Is the understanding that length squared is the unit for the Cosmological Constant a demonstration that the concepts of Euclidean and non-Euclidean geometry are a conflation of geometry and physical theory?

e. Is parallel transport of a vector on a manifold equivalent to all solutions of a constant magnitude for a partial geometric derivative in LSS?

f. In 1998 it was discovered [3,4] that there is unpredicted late inflection point in the expansion (decelerating to accelerating) of the universe. How is this possible at the same moment across the entire universe if nothing can travel faster than light? Would the hypothesis that this is just a property of energy and matter, in that gravitation has a wavelength, be less conceptually offensive?

g. Can we interpret a quantized geometrical wave in LSS as having a wavelength with regions of relative line segments for which $\frac{dS_v^2}{d^2S_h} = nonzero\ constant$ separated by regions for which $\frac{dS_v^2}{d^2S_h} = 0$?

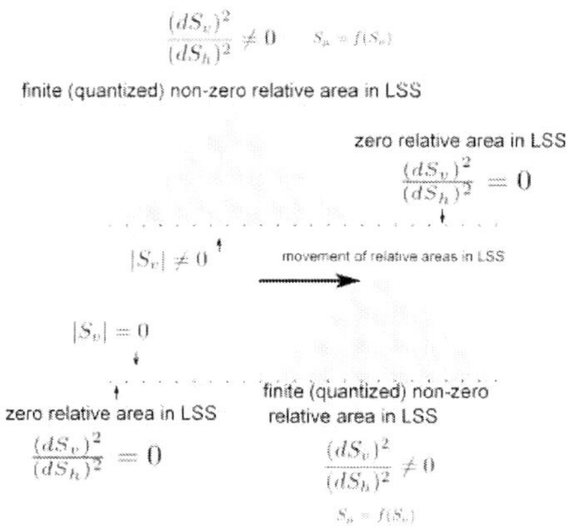

$\frac{(dS_v)^2}{(dS_h)^2} \neq 0$ $S_\mu = f(S_v)$

finite (quantized) non-zero relative area in LSS

zero relative area in LSS

$\frac{(dS_v)^2}{(dS_h)^2} = 0$

$|S_v| \neq 0$

movement of relative areas in LSS

$|S_v| = 0$

zero relative area in LSS

$\frac{(dS_v)^2}{(dS_h)^2} = 0$

finite (quantized) non-zero relative area in LSS

$\frac{(dS_v)^2}{(dS_h)^2} \neq 0$

$S_\mu = f(S_v)$

Figure 9. Proposed quantized waves in LSS. Relative area overlap not shown for clarity.

h. Can Gunnar Nordström's first metric theory [5] be derived from absolute line segments in Line Segment Space?

$$\frac{d^2\phi}{dx_0^2} + \frac{d^2\phi}{dx_1^2} + \frac{d^2\phi}{dx_2^2} + \frac{d^2\phi}{dx_3^2} = i$$

i. Can Gunnar Nordström's second metric [5] theory with variable mass be derived from either absolute or relative line segments in Line Segment Space?

$$\frac{d^2\phi}{dx_0^2} + \frac{d^2\phi}{dx_1^2} + \frac{d^2\phi}{dx_2^2} + \frac{d^2\phi}{dx_3^2} = e^z m$$

j. Can bi-metric gravity theory be derived from using point notation for relative line segments in Line Segment Space where the end points for $S_v = A_1 - A_2$ are conceptually separated into their own line segments so that one can be considered constant while the other is dynamic?

$$ds^2 = g_{\mu v} dx_\mu dx_v$$

$$ds^2 = h_{\mu v} dx_\mu dx_v$$

$$A_1 - 0 \rightarrow \frac{A_1 - 0}{dA_1} = g_{\mu v}$$

$$0 - A_2 \rightarrow \frac{0 - A_2}{dA_1} = h_{\mu v}$$

k. Physical Modeling: Interpretation of the Cosmological Constant

What we find most troubling is that bringing in the concept of the CC as the energy density of the vacuum into LSS would seem to make more physical and equational sense than in GR. Infinitesimal slices of area in IG can be called Geometric One-Forms. For the relative functional relationship $S_\mu = f(S_v)$, it would seem that relative changes in dS_v can be interpreted as dt and dx which are functions of relative changes of S_h (dS_h) which would physically correspond to the relative changes of density and pressure of the vacuum treated as a perfect fluid. Moreover, the one form $S_h dS_v$ would seem to correspond to the integral definition of probability at a point, a foundational assumption of QM.

We consider refuting Integral Geometry worthwhile considering these possible coincidences and the intractableness of the CC problem.

References

[1] Carroll, Sean, 2001, *The Cosmological Constant*, Living Reviews.

[2] Misner, Charles, Thorne, Kip, Wheeler, John, 1973, *Gravitation*.

[3] Perlmutter, Saul et al., 1999, "Measurements of Ω and Λ from 42 high-redshift supernovae", The Astrophysical Journal.

[4] Reiss, Adam et al., 1998, "Observational evidence from supernovae for an accelerating universe and a cosmological constant", The Astronomical Journal.

[5] Renn, Jürgen (Ed.), 2007, *The Genesis of General Relativity*

Unified Field Theory - Finite

Phil Bouchard
20 Poirier St, Gatineau, Quebec, Canada
e-mail: pbouchard8@gmail.com

The current mathematical representation of General Relativity uses a four dimensional reference frame to position in space-time an object with time as a linear variable that can have both a negative and positive value.

In this paper a new mathematical model is suggested based on the classical mechanics, solely on the fact gravity is a particle and that time dilation / contraction is proportional to the kinetic energy and the superposed layers of gravitational potentials. The theory is objective and predicts small scale GPS gravitational time dilation, perihelion precession disparity for all planets, and gravitational light bending. We also consider the rotation curve for all galaxies without dark matter, the natural faster-than-light expansion of the universe without dark energy, even the constitution of a black hole and the velocity of the visible universe. In particle terms since gravity can be manipulated, the theory allows time travel into the future, infinite speed and levitation.

1. Introduction

Nomenclature:

FT: Finite Theory	GR: General Relativity

FT defines a new representation of the formulas derived from GR based on the superposed potentials of the predicted massless spin-2 gravitons that mediate gravitational fields.

Additionally in contrast to GR where the space-time is represented using the non-Euclidean geometry in order to keep the speed of light constant, FT considers time to be a positive variable within a space that is characterized by the Euclidean geometry. No previous self-consistent results deriving from GR are in violation.

FT postulates time dilation to be directly proportional to its energy, which is later shown to be sufficient to explain all anomalies:

1. The kinetic energy of body relative to its maxima induces dilation of time
2. A gravitational time dilation is the direct cause of the superposed gravitational potentials

1.1 Black Hole Radius

The Schwarzschild radius defines the event horizon where the gravitational pull exceeds the escape velocity of the speed of light. This is given by:

$$r_s = \frac{2GM}{c^2} \quad (1)$$

Given that Schwarzschild radius derives from GR formulation, FT will need its own definition. Satisfyingly, this event horizon can easily be found with the amount of kinetic energy needed to overtake the gravitational potential energy:

$$\frac{1}{2}mv^2 = \frac{GMm}{r_b} \quad (2)$$

By solving the equation with the maximum escape velocity a photon can have, where the mass is of non-importance we get:

$$r_b = \frac{2GM}{c^2} \quad (3)$$

Despite the fact the resulting equation is exactly the same as the Schwarzschild radius, we will use a different notation given that its origin differs.

1.2 Gravitational Time Contraction

Gravitational time contraction will be used interdependently with the non-trivial superposed gravity field of the observer and the observed object. As a result the potentials are fractionalized in order to calculate the conversion factor.

1.2.1 Outside a Sphere

Since an inertial body being subject to a specific gravitational force is responsible for gravitational time dilation and that gravity is a superposable force, we will translate the same conditions of all gravitational potentials into the sum of all surrounding fields of an observed clock and the observer:

$$t_o = \frac{\Phi(r)}{\Phi(r_o)} \times t_f \tag{4}$$

$$t_o = \frac{\sum_{i=1}^{n} \frac{m_i}{|r_i - r|}}{\sum_{i=1}^{n} \frac{m_i}{|r_i - r_o|}} \times t_f \tag{5}$$

Where:

- r is the location of the observed clock
- r_i is the location of the center of mass i
- r_o is the location of the observer (typically 0)
- m_i is the mass i
- t_o is the observed time of two events from the clock
- t_f is the coordinate time between two events relative to the clock

By juxtaposing the same spherical mass with its external gravitational time dilation factor and internal counterpart we have the following, for a spherical mass of 20 meters in radius:

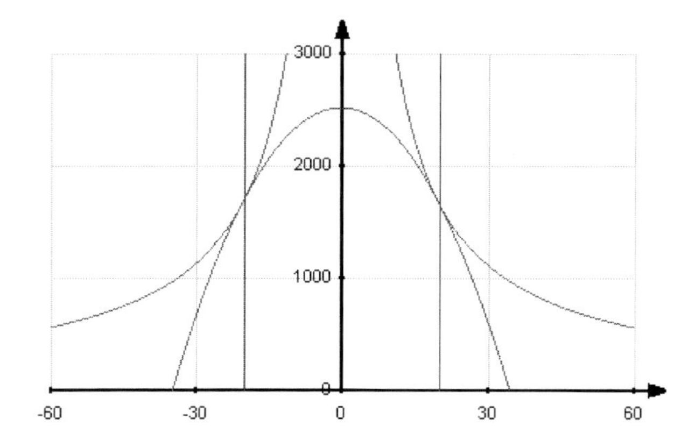

Fig. 1. Inner & Outer Gravitational Time Dilation Factors vs. Radius *(m)*

2 Implications

Herein are enumerated all consequences FT will lead to and important differences from GR are highlighted. No precise mathematical proof is being made in this matter; only logical observation, deductions and estimates are necessary to disjoint many hypotheses.

At this level only complex computer research can be proposed to simulate a modelling of the universe under this umbrella in order to match its behaviour with measurements such as the constant of the Hubble's Law. Potentially, simulators can also be used to reverse time and estimate an early universe according to the current velocities of the superclusters, solve the scaling factor of the observed universe which will lead to an estimation of the real volume of the universe and solve local focal points of gravitational lenses.

2.1 Fallacy of General Relativity

If we equate equations (9) and (11) by using a reference point infinitely far away and letting h include the effects of the Sun:

$$\frac{\frac{m}{r} + h}{h} = \frac{\sqrt{1 - \frac{2Gm}{rc^2}}}{1} \tag{6}$$

$$\frac{m}{rh} + 1 = 1 + \frac{Gm}{rc^2} \tag{7}$$

We will observe that General Relativity is making use of a constant in its equations:

$$h = \frac{c^2}{G} \tag{8}$$

As it turns out, this happens to follow the same ratio used by Einstein's constant:

$$\kappa = \frac{-8\pi G}{c^2} \tag{9}$$

Thus General Relativity is not scalable.

2.2 Artificial Faster-Than-Light Speed

By creating a tunnel with a lower gravitational potential we will observe beams of light traveling faster than c for an observer outside of the tunnel. In the following case we reach infinite speed for an object in a tunnel approaching a null gravitational potential:

$$v_o = \lim_{\Phi(r)\to 0} \frac{\Phi(r_o)}{\Phi(r)} \times v_f \tag{10}$$

$$v_o = \infty \tag{11}$$

2.3 Dark Matter and the Galactic Rotation Curve

The idea of dark matter [1] is supposed to replace the missing matter necessary to withhold all tangential galaxies within their cluster traveling much higher than the necessary escape velocity. Dark matter explains also the same scenario at lower scales where tangential stars should technically easily escape the attraction towards to centre of their galaxy. Unfortunately after many attempts of unfolding the nature of dark matter, no conclusive discovery can be revealed.

In contrast, by using FT as a mathematical representation we will find much different conclusions. Indeed, the stars and galaxies rotating around their galaxy and cluster respectively will be subject to time contraction. This means the bodies will be seen to travel much faster than the anticipated Newtonian speed. There is therefore no need for any dark matter to increase the gravity strength necessary to keep the tangential objects in an uninterrupted cycle.

In the other hand, if we add time contraction effects to the stars orbiting the galaxy we will get very different results. Let's imagine our neighbour Andromeda has exactly the same properties as the Milky Way, in order to simplify our measurements, and we are observing it from our solar system. In these conditions an approximation of the observed speed of the rotating stars of Andromeda as seen from our position can be given by the following according to FT:

$$v = \sqrt{\frac{Gm}{x} \times \left(\frac{m}{r} + h\right) \Big/ \left(\frac{m}{x} + h\right)} \tag{12}$$

Where:

- $m = 1.1535736 \times 10^{42}\ kg$
- $r = 2.45986 \times 10^{20}\ m$
- $h = 2.5 \times 10^{22}\ kg/m$

We have arbitrarily adjusted the scaling factor h of the Virgo cluster properly to show the effects on the subjected Milky Way galaxy:

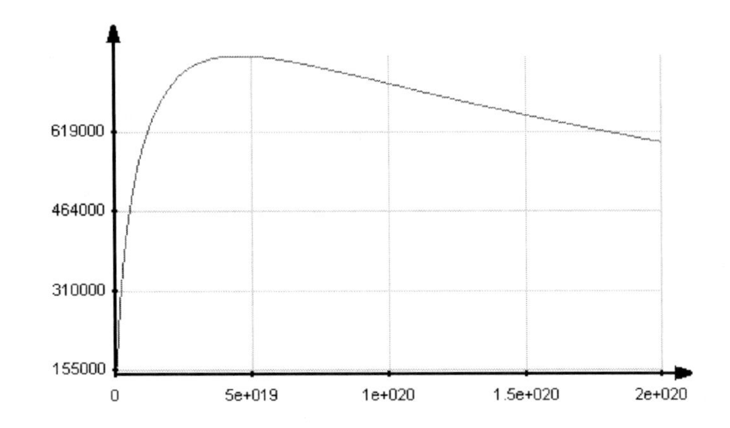

Fig. 2. Orbital Velocity (m/s) vs. Radius (m)

We clearly see the observed velocity of the stars in Andromeda with different radius than our own Sun in the Milky Way ($r_o \neq r$). The graph curve is consistent with what is currently observed with distant galaxies.

The aforementioned rotation curve matches most of the galaxies, however low surface brightness galaxies have shown a much different trend [2]. Indeed the galaxies in question indicate an extremely high mass-to-light ratio, which will consequently affect the observed rotation curve. In the context of FT this can be accomplished by simply lowering its scaling factor:

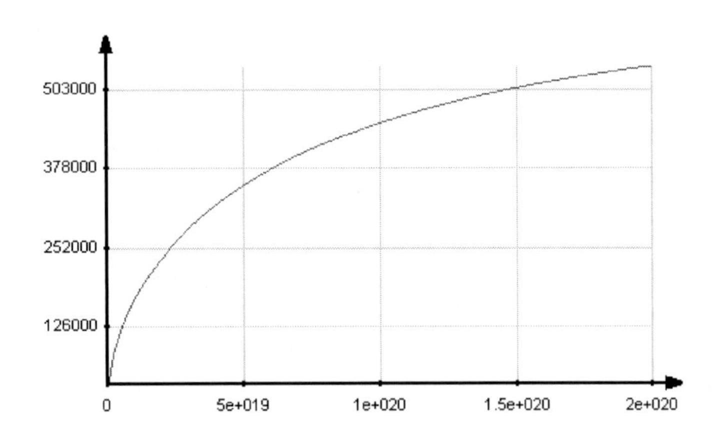

Fig. 3. Orbital Velocity (m/s) vs. Radius (m)

Where:

- $h = 2.5 \times 10^{21}\ kg/m$

2.4 Dark Energy and the Centre of the Universe

Dark energy is a constant or scalar field filling all of space that has been hypothesized but remains undetected in laboratories [3].

The problem is that in order to do so the amount of vacuum energy required to overcome gravitational attraction would require a constantly increasing total energy of the universe in violation of the law of conservation. The Hubble's law represents the rate of the expansion of the universe with the speed of the distant galaxies as seen from the Milky Way with:

$$v_o = H_0 x \tag{13}$$

Where:

- $H_0 = 2.26\times10^{-18}\ s^{-1}$ (Hubble's constant)

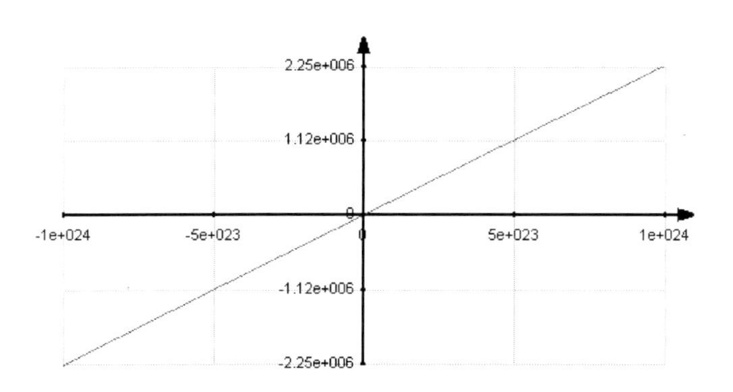

Fig. 4. Speed (m/s) vs. Distance (m)

On the other hand FT applied on the scale of the universe proves that there is no need for such energy. Indeed if we consider the universe to be the result of a Big Bang then all galaxies must have a certain momentum. If we try to represent the speed of the observed galaxies using FT where h is null because the environment must not be encompassed by anything else then we will have:

$$v_o = \frac{\frac{m}{|i|}}{\frac{m}{|x-i|}} \times v_f \tag{14}$$

After simplifying, subtracting the speed of the observer from his observations and disposing the absolute values to keep track of the direction we will have:

$$v_o = \frac{i-x}{i} \times v_f - v_f \tag{15}$$

Where:

- $i = -2.66\times10^{23}\ m$ (position of the kernel)
- $v_f = 6\times10^5\ m/s$ (speed of the observer)

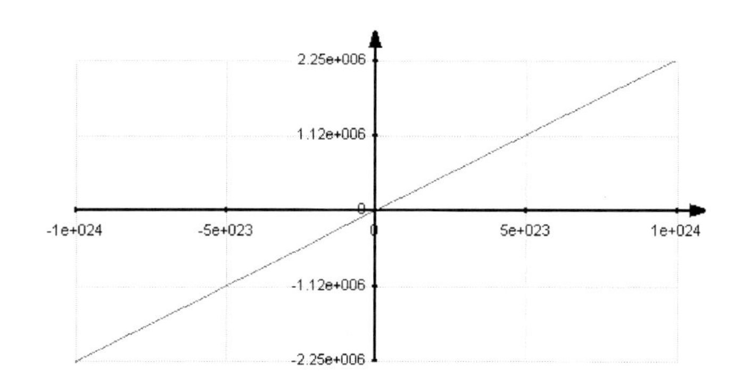

Fig. 5. Speed (m/s) vs. Distance (m)

This means i, or the position of the centre of the universe is actually solvable by equalling Equ. *(13)* and *(15)*:

$$H_0 x = \frac{i-x}{i} \times v_f - v_f \tag{16}$$

$$i = -\frac{v_f}{H_0} \tag{17}$$

The speed of the observer v_f is actually that of the Milky Way but the visible universe itself most likely has its own momentum so the latter has to be taken into account.

3. Conclusion

It was shown that every single observation at the galactic and the universe scales can be explained with the effects of the gravitational time dilation / contraction.

The same is also true at the solar system scale with the classic perihelion precession disparity and the gravitational light bending as estimated by a computer simulator.

Even accelerated particles are covered by the theory as the kinetic time dilation is actually responsible to alter the momentum of the particle.

This theory is deductive and replaces GR at all scales without the need for any dark matter and dark energy. The unknown now resides at scales greater than the visible universe with no obstacle in between.

Acknowledgments

This article was still a theoretical facet in the year 2005 and was fostered into a serious labour after confirmation of its potential validity by Dr. Griest and important requirements. Many thanks to him.

Furthermore my father, Mr. Bouchard M.Sc. Physics, brought considerable help in asserting the mechanical part of my equations. He also introduced me to the astrophysical society for advanced research.

The same goes directly and indirectly for the online scientific community, where we can find Peter Webb, Greg Neill and Paul Draper. Also thanks to Peter Watson, Jim Black and Robert Higgins for a fix of the maximum mass density formula, an inside the sphere calculation error and textual corrections respectively. We can also find the great resources on Cosmoquest.org where I have debate it.

Finally thanks to Dimtcho Dimov some help solving an equation and Evan Adams for helping editing it and for reviewing related experiments.

References

[1] White, Nicholas. Beyond Einstein: From the Big Bang to Black Holes. http://www.phy.bnl.gov/~partsem/fy07/Brookhaven%20nwhite.pdf

[2] De Blok, W. J. G. Halo mass profiles and low surface brightness galaxy rotation curves. IOP Institute of Physics. http://iopscience.iop.org/0004-637X/634/1/227/pdf/62338.web.pdf

[3] Correa, Paulo, and Correa, Alexandra. What Is Dark Energy. http://www.aetherometry.com/

The Mediums for Light are Hiding in Plain Sight

Richard O. Calkins
2125 Sahalee Dr. E., Sammamish, WA, USA
Email: rocalkins@msn.com

The propagation of light remains one of the enduring mysteries of science. Unlike every other known kind of wave, it travels at a constant speed through empty space without a medium of propagation. What supports its travel in empty space? Why is its speed so constant? And why is it so very fast? Even Einstein gave up trying to puzzle it out and simply stated the propagation of light as a postulate. While Maxwell's equations explain the behavior (the what) of light, they don't explain the how or the why. They show that the speed of light is determined by two constants - ϵ_0 and μ_0 - but not why these two constants have the values that they do rather than having some other values. So our questions remain: how does light propagate through empty space and why do ϵ_0 and μ_0 have the values that they do? Maxwell's equations actually do answer these questions. We just haven't understood the message because of our human laundry list of implicit assumptions about the characteristics required of mediums of propagation. As often happens with implicit assumptions, nearly all of them are wrong.

Light has not one but two mediums of propagation. The electric fields in electromagnetic waves are the medium of propagation for the magnetic fields, and the magnetic fields are the medium of propagation for the electric fields. Their dance with each other is what moves light through empty space, much like molecules of air do when propagating sound. Ironically, since the waves of light and the dancing electric and magnetic fields are the same things, the mediums of propagation quite literally have been hiding in plain sight.

This paper examines both the equation for the speed of sound and Maxwell's equations for the propagation of light from a new perspective. In essence, it performs a mathematical autopsy of both to illuminate what lies within and, in the process, solves the mysteries of light's propagation as simply the laws of physics at work.

1. Introduction

The Michelson-Morley experiment is universally accepted as conclusive proof that light races through empty space at a definite speed c=299,792.5 km/s in the total absence of any medium of propagation whatever.i However, all that their experiment actually proves is that there is no stationary medium permeating empty space through which light waves propagate. That by no means rules out a medium of propagation which operates under different laws of physics than those of more familiar kinds of waves.

The design of the Michelson-Morley experiment and the interpretation of its empirical observations were governed by a number of implicit assumptions about the innate nature of mediums of propagation. These implicit assumptions strongly influenced the interpretation of the Michelson-Morley experiment. That influence remains to this day. But as is the case with implicit assumptions, they have not been subjected to rigorous conscious examination. And, in the case of light waves, nearly all of them are wrong. These implicit assumptions include:

- A wave's medium of propagation is physically separate from the wave.[ii]
- The medium of propagation must permeate wherever the wave is able to travel.[iii]
- The medium of propagation is present whether or not the wave is passing through it.[iv]
- The medium of propagation is stationary and the wave travels through it at its speed of propagation.[v]
- The speed at which the wave travels is determined by the physical characteristics of its medium of propagation and the laws of physics which govern them.[vi]

In the case of light waves, only the last assumption is correct. All of the first four are wrong. But our subconscious, unexamined faith in them is what has blinded us from recognizing light's mediums of propagation even though they are clearly described by Maxwell's equations (another example of the human condition at work).

As Maxwell's equations tell us, light has two mediums of propagation, its constituent electric and magnetic fields. A light wave and its mediums of propagation are the same physical entities. The mediums are not stationary. Because the wave and the mediums are the same entities, they move through empty space together. Thus, they only need to be where the wave is; they do not have to permeate the space through which it travels. They are present only when the wave is present. And the wave does not propagate "through" them; they move with the wave. In fact, they are the wave and their dance with each other is both what propagates the wave and determines its speed. Only the last implicit assumption applies to light. Its speed of propagation is determined by the physical behavior of electric and magnetic fields and by the laws of physics which govern that behavior.

2. Maxwell's Equations

Maxwell's equations begin by describing the original source of an electric field; the presence in space of an electric charge.vii Maxwell's first equation:

$$\nabla \cdot E = \frac{\rho_{EC}}{\epsilon_0}$$ viii

says that the amount of electric field (E) that is coming from a location in space ($\nabla \cdot$) is equal to the density of the electric charge at that location (ρ_{EC}) divided by a constant (ϵ_0). For the moment, we can think of ϵ_0 as a fudge factor to make a proportional relationship between $\nabla \cdot E$ and ρ_{EC} into full equality. This allows us to use the relationship in mathematical calculations with greater precision than when limited to mere proportionality. The name for this kind of fudge factor is "proportionality constant." Like most proportionality constants, its value was determined by experiment. That means someone made careful measurements to figure out what its value had to be to make the two sides of the equation equal.

To better understand Maxwell's first equation, it may be helpful to examine the concepts of electric charge and electric field. Electric charge is an innate characteristic of certain subatomic particles. Protons are tiny particles that usually reside in the nucleus of an atom. Electrons are even tinier particles that orbit the nucleus. The electric charges of the two particles are equal in magnitude but opposite in their behavior. Protons attract electrons and repel protons. Electrons repel electrons and attract protons. Because they were such opposites, protons were declared to have a positive charge and electrons to have a negative charge. (Apparently, protons had better lobbyists than electrons.) However, the + and - designations for the two opposite charges actually turn out to be useful. For example, if we catch 10 electrons and 7 protons and put them into a box, the amount and sign of the box's charge would be:

$$\text{Net charge} = -10 + 7 = -3$$

The electrical behavior of the charge in the box would be the same as if we couldn't catch any protons and just threw three electrons into it. Thus, we can think of electric charge both at the particle level and at the level of larger bodies that have either more or fewer electrons than they do protons. In either case, the strength of an electric charge to do things, like push other electric charges around, diminishes as you move away from its source. Thus, electric charge is what's called a point charge. It exists at the location of its origin. The farther you go from that location, the less its presence. However, it still does have a presence, even though diminished.

In the early work on electric charges, their ability to do things at a distance was troubling to those doing the experiments. In mechanical systems, one body can impose a force on another only when in physical contact with it. Electric charges were not limited in that manner. To deal with this effect at a distance, the British scientist Michael Faraday introduced the concept of an electric field. As shown in Figure 1, an electric field extends outward in all directions from the location of the charge. In Faraday's field

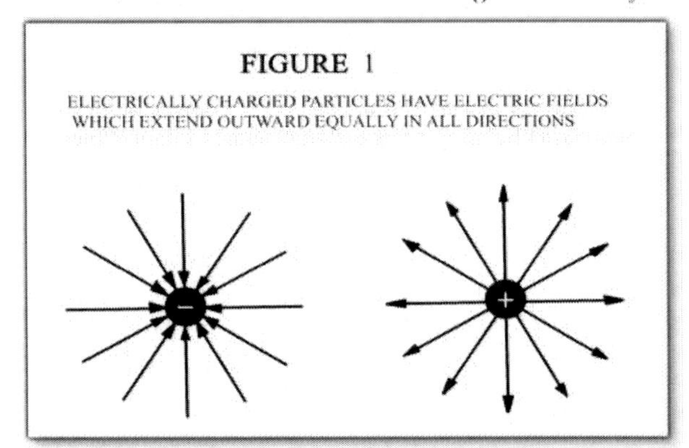

concept, it is the electric field of one charged entity that pushes or pulls on the electric field of another. If the fields are unlike each other, they will attract each other. But as shown in Figure 2, if the fields are the same as each other, they will push against each other. However, as the distance from a charged entity increases, its electric field is distributed over a greater area. Thus, its force is diminished. Because the field expands outward from the charged particle equally in all directions, the area over which it is spread at a distance r is equal to the area of a sphere whose radius is r. The equation for the area of a sphere is:

$$A = 4\pi r^2$$

Thus, if an entity whose electric charge is Q has an electric field E when measured directly at its location in space, the strength (or amount) of electric field E_r measured at a distance r should be equal to E divided by the area of a sphere with radius r. However, what will be measured at that distance is not what one would expect from the increased area over which the field is spread. Instead of being equal to E divided by $4\pi r^2$, the value of E_r is proportional to the expected result but not equal to it.

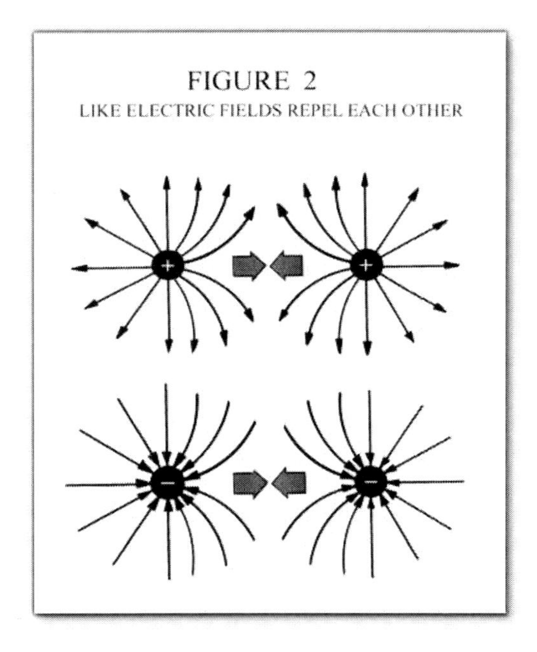
FIGURE 2
LIKE ELECTRIC FIELDS REPEL EACH OTHER

$E_r \propto \frac{E}{A}$ And $A = 4\pi r^2$, which means

$$E_r \propto \frac{E}{4\pi r^2} \quad \text{Proportionality 1}$$

Why is E_r merely proportional to E divided by $4\pi r^2$ rather than being equal to it? The only difference between them is that E is measured at the source and E_r is measured at the distance r from the source. That spreads E over a larger area; the area of a sphere with radius r and a surface area of $4\pi r^2$. What else is going on?

As it happens, the French physicist Charles Coulomb made extensive experiments of the interactions between charged entities in the 1780s. The results of his experiments are expressed in Coulomb's law for the amount of electric field E_r resulting from a single point charge Q at a distance r:

$$E_r = \frac{1}{4\pi\epsilon_0}\frac{Q}{r^2} \quad \text{Equation 2 (Coulomb's law)}$$ [ix]

We can rearrange Coulomb's law for comparison with Proportionality 1 as follows:

$$E_r = \frac{Q}{4\pi r^2}\frac{1}{\epsilon_0} \quad \text{(another form of Coulomb's law)}$$

Equation 3

Comparing Equation 3 with Proportionality 1, we can see that this is exactly what we would expect at distance r, x except that we must include that same constant ϵ_0 that we found in Maxwell's first equation to get from proportionality to equality. Comparing Equation 3 with Maxwell's first equation shows that the density of an electric charge (ρ_{EC}) at a distance (r) from its source is equal to the charge at its source Q adjusted for the area over which it is spread at that distance:

$$\rho_{EC} = \frac{Q}{4\pi r^2}$$

Substituting this value for ρ_{EC} in Maxwell's first equation produces the same equation as Coulomb's law.

$$\nabla \cdot E = \frac{\rho_{EC}}{\epsilon_0} = \frac{Q/4\pi r^2}{\epsilon_0} = \frac{1}{4\pi\epsilon_0}\frac{Q}{r^2} \quad \text{Equation 4}$$

Maxwell's first equation   Coulomb's law

Maxwell's first equation and Coulomb's law are two ways of saying the same thing. $\nabla \cdot E$, in Maxwell's first equation, is just another way to say E_r, except it goes on to say that the amount of electric field is coming *from* the density of the electric charge at that location rather than merely being equal to it. The difference between "coming from" and "equal to" reflects Maxwell's intuition that the force applied by an electric field is dynamic. There is something more involved than simply being present. (e.g., A teacher must do more than merely be present to pass knowledge to students. There must be some form of interaction as well.)

Why did we go through this? To show that Maxwell's first equation, taken in context with Coulomb's law, tells us two important things. First, the force of a point charge's electric field spreads out equally in all directions. Thus, it will diminish with distance at a rate which is proportional to the increase in the area over which it is spread. Second, something additional is going on that is related to the constant, ϵ_0. Clearly, ϵ_0 has something to do with the fact that an electric field must act dynamically in order to be expressed. It can't express itself merely by being present.

Now let's see what we can learn from Maxwell's second equation:

$$\nabla \cdot B = 0.$$

$\nabla \cdot B$ stands for the amount of magnetic field B coming from a location in space. The equation tells us that it is equal to … zero? Actually, what that means is that magnetic fields don't come from a point charge that has a location in space. Wherever you find a magnetic field the whole field is present and accounted for. Magnetic fields always go around in loops. They don't have a beginning or an end. (Note that the symbol B in this equation stands for the magnetic field and not for a bulk modulus. Every use of B for a bulk modulus in this report has a subscript to denote the medium to which it belongs.)

Maxwell's third equation:

$$\nabla \times E = -\frac{\partial B}{\partial t}$$

tells us that there is another source for an electric field beyond the one described in Maxwell's first equation. The electric field described in Maxwell's first equation comes from an electrically charged entity that exists at a point in space. This electric field is caused by the movement of a magnetic field. What we need to know here is that the symbol $\nabla \times E$ stands for the curl of the electric field. The term "curl" here means just what it says. *This* electric field goes around in loops, just like the magnetic field always does. $\nabla \times E$ is minus the rate of change of the magnetic field $\partial B/\partial t$. (The minus just means that E curls in the opposite direction from B.) As shown in Figure 3, increasing the amount of curl means reducing the radius of the arc. Thus, the curlier the field, the tighter its circles are wrapped around each other and the greater is the field's density. This simply means that the faster the magnetic field changes, the

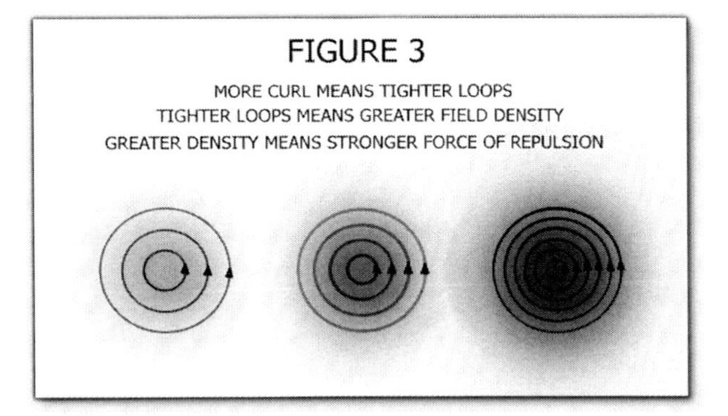

denser is the electric field it creates. The denser the field, the more energy it has packed into it and the greater the strength of its compulsion to dissipate and of its opposition to further compression. That is the same behavior as that of air. The more densely the air is compressed, the harder it pushes to expand and the harder it is to compress.

Maxwell's fourth equation:

$$\nabla \times B = \mu_0 J + \mu_0 \epsilon_0 \frac{\partial E}{\partial t}$$

says that the curl (field density, field strength) of the magnetic field is determined by the sum of two things, both of them related to the movement of electric fields. The first is the density (amount) of an electric current J multiplied by another constant μ_0. The second is the rate of change in the electric field ($\partial E/\partial t$) multiplied by both constants μ_0 and ϵ_0. The specific values of those two constants were determined by what they had to be to make the results of the equations match the values observed in the experiments. As it turned out, they had quite a story to tell.

That story begins with Maxwell's recognition that his fourth equation shows you can have a magnetic field without having an electric current. If you set J to zero, his fourth equation effectively becomes:

$$\nabla \times B = \mu_0 \epsilon_0 \frac{\partial E}{\partial t} \quad \text{Equation 5}$$

No current is present, but we have a magnetic field as long as there is a changing electric field. And remember that Maxwell's third equation shows that all you need in order to have an electric field is a changing magnetic field. Both the electric and magnetic fields are curled into loops rather than emanating outward from a point source. Because each type of field in motion creates the other, they can go on supporting mutual creation indefinitely. (We call it electromagnetic radiation.)xi

Taking his third equation and his fourth equation with J set to zero (Equation 5), Maxwell developed two more equations that tell us how the electric and magnetic fields move in space as a result of their change in time:

$$\nabla^2 E = -\mu_0 \epsilon_0 \frac{\partial^2 E}{\partial t^2} \quad \text{Equation 6}$$

$$\nabla^2 B = -\mu_0 \epsilon_0 \frac{\partial^2 B}{\partial t^2} \quad \text{Equation 7}$$

On the right hand side of each equation is the change in the field's rate of change in time and on the left hand side is the change in the same type of field's rate of change in space. An electromagnetic wave's speed of propagation turns out to be dependent on the values of μ_0 and ϵ_0. Maxwell solved these differential equations to determine that speed as:

$$v = \frac{1}{\sqrt{\epsilon_0 \mu_0}}$$

When he plugged the experimentally derived values for μ_0 and ϵ_0 into this equation, the value of v turned out to be the same as the previously determined speed of light. This is how Maxwell discovered that light is an electromagnetic wave. Since the speed of light in empty space always is precisely the same, it has been assigned its own symbol c and the equation for its speed is

$$c = \frac{1}{\sqrt{\epsilon_0 \mu_0}}$$

Note that this equation also says that the speed in the vacuum of empty space of every electromagnetic wave, regardless of wavelength, frequency or amplitude, is precisely the same. Also, because the equation contains only two empirically derived constants, this also strongly (but incorrectly) suggests that the speed of light (in its broad sense as the speed of all electromagnetic waves) is independent of physical laws. After all, a constant is just a constant.

This segue through Maxwell's equations was made to develop an understanding of how the determinants of the speed of light compare with those of the speed of sound. But before we do that, it's worth noting some of the implications and interpretations about electromagnetism that have resulted from the structure of Maxwell's equations:

First, the structures of Equations 6 and 7 are identical.

$$\nabla^2 \quad = -\mu_0 \epsilon_0 \frac{\partial^2}{\partial t^2}$$

If you put the electric field in the box on the right, you get the rate of change of the electric field's rate of displacement in space on the left. If you put the magnetic field in the box on the right, you get the rate of change of the magnetic field's rate of displacement in space on the left. This appears to have caused a perception that electric and magnetic fields might be coequal. There have been several attempts to identify a magnetic monopole, so far without success. However, this coequality is only apparent, not real. An electric field can exist in the absence of a magnetic field. Its original source is a charged particle (Maxwell's first equation). A magnetic field exists only when being created by an electric field in motion. The appearance of equality comes from having removed the electric current (J) from Maxwell's fourth equation (producing Equation 5) before developing Equations 6 and 7. When the electric current is removed, the electric field is reduced to the same dependency on the magnetic field as the magnetic field always has on the electric field. Once an electromagnetic wave leaves its source, the only electric field it contains is a loop-shaped (not point-sourced) electric field created by a moving magnetic field. This co-dependency between the motions of the two fields in an electromagnetic wave is why it can be said that when a photon stops moving it ceases to exist.

Another interpretational problem that arises from the structure of equations 6 and 7 is that each field's change in displacement is described in terms of its own change in time. That implies, incorrectly, that each field somehow moves itself through empty space. That, in turn, omits something really important that happens in between. An electric field that changes in time does not directly create an electric field that moves in space. According to Equation 5, what it does is create a magnetic field. According to Maxwell's third equation, that magnetic field's motion then creates the next electric field. Once begun, this interaction between the two types of field will continue until interrupted by an outside force. By failing to express these essential interim steps, equations 6 and 7 lull us into thinking that the medium of propagation is the vacuum of free space. Not so. The medium of propagation for the moving electric field is the magnetic field it must push into existence as an unavoidable consequence of its movement. The magnetic field starts with zero density (it doesn't yet exist) and moves to greater density as the moving electric field pushes and curls it into existence. It is in the nature of the magnetic field to resist having its density increased. Just like air, the magnetic field being pushed into existence has both a density and an innate resistance to being compressed (a.k.a. a bulk modulus). It is inarguable that the magnetic field is acting as the medium of propagation for the electric field. It is actively created by the motion of the electric field and it is the only active element in the creation of the next electric field. The same is true of the electric field's characteristics and its actions to propagate the magnetic field.xii The same phenomena are at work in a similar manner for the propagation of light as for the propagation of sound. A disturbance in the electric field's density initiates an ongoing interaction between its compulsion to expand at one area in space and the magnetic field's opposition to compression in the adjacent area in space. The magnetic field, being on the opposite side from the electric field on each transaction, takes turns in performing the same functions for the electric field as the electric field performs for the magnetic field. In the case of sound, there is only one medium on both sides of the interaction between the medium's compulsion to expand and its opposition to compression. Light has two mediums. Whichever field is dominant (i.e., the more powerfully compressed) can decompress only by overcoming the other field's resistance to compression. The electric and magnetic fields simply take turns as they work their alternate sides of the transaction.

What we failed to realize when we accepted μ_0 and ϵ_0 as simple constants, derived to make the equations meet the observations, is their underlying physical significance.xiii ϵ_0 is not the "permittivity of free space." It has nothing to do with the characteristics of space. It is determined by how hard an electric field pushes against itself. That, in turn, determines how fast it will move, in order to decompress, when its motion is unopposed. Even when its motion is unopposed, that speed of decompression is not infinite. It is a function of the strength of the innate force of repulsion of like electric fields (a.k.a. a function of the electric field's bulk modulus). The same is true of μ_0 for a magnetic field. This will be explained more fully in Section 4. But before we do that, we must become more familiar with the mechanics of how air's response to a sound wave causes it to act as the medium of propagation for sound. The equation for the speed of sound tells us what determines the speed of sound, but does not explain how it determines the speed of sound.

3. Dissecting The Equation For The Speed Of Sound

On the surface, the equations for the speed of sound and the speed of light are totally different from each other. Using v_s as the symbol for the speed of sound, the equation for the speed of sound is:

$$v_s = \sqrt{\frac{B_a}{\rho_a}} \qquad \text{Equation 8}^{\text{xiv}}$$

In the equation, B_a stands for the bulk modulus of air and ρ_a stands for its density. The bulk modulus describes air's innate compulsion to expand and, on the other side of the same coin, its innate resistance to compression. The formula for air's bulk modulus is:

$$B_a = -\frac{\Delta p}{\Delta V/V_0} \qquad \text{Equation 9}^{\text{xv}}$$

As shown in Equation 9, the value of B_a is determined by the change in pressure Δp that is required to reduce the volume by a given amount ΔV relative to the initial volume V_0. (The minus sign just means that the pressure and volume change in opposite directions. When pressure is increased, volume is reduced, and vice versa. The result is always a positive number for B_a). The more pressure that is required to produce a given reduction in volume (i.e., the harder it is to compress the medium), the greater the value of B_a and, as shown in Equation 8, the faster the wave will move. ρ_a is the density of air. The greater the density, the slower the wave will move. These two characteristics of air are what determine the speed of sound. However, to compare Equation 8 with the equation for the speed of light, it will be necessary to look inside it for more information. So … get out the scalpel.

Instead of focusing on what the speed of sound v_s is equal to in equation 8, one must focus on what B_a/ρ_a is equal to.

$$\frac{B_a}{\rho_a} = v_s{}^2 \quad \text{Equation 10}$$

Looking at it that way raises an interesting question: Why is the ratio of air's compressive-resistance B_a to its density ρ_a equal to the square of the speed of sound v_s? Why isn't it directly equal to the speed of sound? What is going on inside of B_a/ρ_a that makes it equal to v_s^2? Note that the standard equation for velocity is

$$v = \frac{d}{t} \quad \text{Equation 11}$$

Velocity is defined in terms of distance traveled per unit of time. Thus, if $B_a/\rho_a = v_s^2$, one also must conclude that:

$$\frac{B_a}{\rho_a} = (d/t)^2 = \frac{d}{t}\frac{d}{t} \quad \text{Equation 12}$$

Thus, the ratio of B_a to ρ_a can be considered as being equal to the product of two velocities. However, one must take care not to fall into the trap of an implicit assumption. In order for v_s to be equal to the square root of B_a/ρ_a it is not necessary for the two velocities whose product is equal to B_a/ρ_a to each be equal to v_s. For example, suppose there are two velocities $v_1 = d_1/t$ and $v_2 = d_2/t$.

$$\text{If} \quad \frac{d_1}{t}\frac{d_2}{t} = \frac{B_a}{\rho_a} \quad \text{then} \quad v_s = \sqrt{\frac{B_a}{\rho_a}} = \sqrt{\frac{d_1}{t}\frac{d_2}{t}}$$

Equation 13

Thus, in order to more fully understand what determines the speed of sound one must understand the nature of the two velocities whose product is equal to the ratio of B_a to ρ_a. In order to do that, it will be useful to recognize that there are two ways of defining the rate of movement, one way is velocity and the other is its reciprocal.

$$v = \frac{d}{t} \quad \text{and its reciprocal} \quad \omega = \frac{1}{v} = \frac{t}{d}$$

v and ω are simply two ways of saying the same thing. One way, velocity, focuses on the ratio of distance to time. The other focuses on the ratio of time to distance. The former is more easily recognized simply because we are almost always interested in how fast an identifiable object is moving. But for expressing the motion of something as amorphous as air pressure or the pressure of the self-repellence of like electric and like magnetic fields, the latter can be more useful. A pocket of pressurized air and an electric or magnetic field exist over an area. If the field is in motion, the entire field is in motion. If one is measuring a field "coming from" somewhere (e.g., as in Maxwell's equations) then it is more useful to think of it as a rate of flow. The amount one receives of whatever is flowing to him is determined by how much stuff there is and how long it takes to be delivered. The time it takes to be delivered is the reciprocal of how fast it is moving.

A sound wave is simply a pocket of compressed air whose pressure moves through a sea of less compressed air. The molecules of air only move back and forth. It is the pocket of pressure which moves through the sea of air from source to listener. The movement of that pocket of pressure results from the interplay between the innate compulsion of the pressurized pocket of air to reduce its pressure by expanding and the innate resistance of the surrounding air to being compressed. To understand what happens, one must consider that it involves pressures on both sides of the transaction, each trying to create motion, but in opposite directions. The forces are pushing against each other. Either force, acting alone, would determine the velocity at which the air on its side of the transaction would move. Their net effect is to determine the speed at which the higher-pressure side expands at the expense of the other side. Whichever medium on one side of the traction is expanding, the medium on the other side is being compressed. Also, each medium's pressure to expand at any given level of compression is the same as its pressure to resist compression. The same forces are at work at the same level of intensity in both directions. Thus, one can think of rate at which a domain of medium can expand as being the same as the rate by which it can oppose compression.

The first velocity to address for the propagation of sound is the speed at which a pressurized domain of air can expand to relieve its pressure when its expansion is unopposed. As shown in Figure 4, if a ball of pressurized air were to be released instantly in empty space, it would be free to expand outward in all directions without anything to constrain it. The speed at which it can expand is determined only by the innate intensity of its self-opposition. One can define that speed as the maximum rate at which the air pressure can expand when unopposed. Its rate of motion can

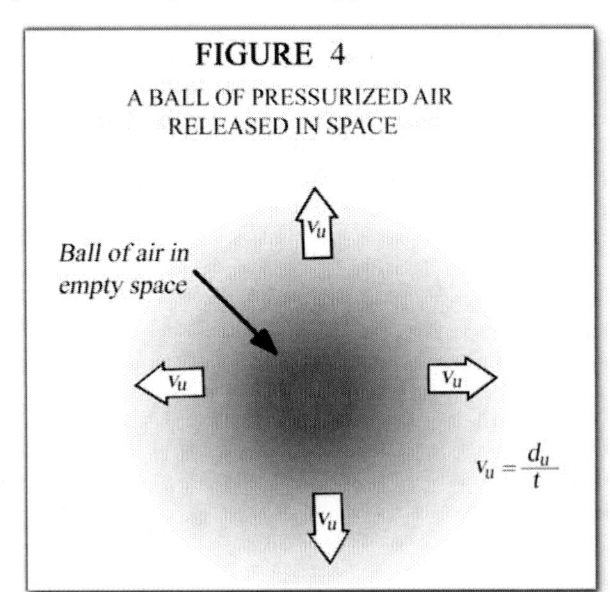

be expressed either as its velocity or how long it takes for it to move.

$$v_u = \frac{d_u}{t} \quad \text{or} \quad \omega_u = \frac{t_u}{d}$$

We don't have to know the precise magnitude of v_u to understand the function it plays in determining the speed of sound. We know its value is a constant which is greater than zero and smaller than infinity. Otherwise, the speed of sound would be either zero or infinite. We also know from equation 12 that the square root of its product with another velocity (one which acts as an environmental constraint) determines the speed of sound. That is all we need to know to understand its function in determining the speed of sound.

When the pocket of pressurized air is a sound wave on Earth, it will be surrounded by ambient air. Thus, the other source of velocity is encountered in the form of a constraint; the resistance to compression of the ambient air. But as discussed above, the intensity to expand for any given domain of medium is identical to the intensity with which it opposes compression. If a pocket of the ambient air were suddenly released in empty space, the velocity at which it would expand would be:

$$v_o = \frac{d_o}{t}$$

But since the ambient air acts as a constraint on the sound wave's expansion, it is best understood in the time constraint form of the motion ω.

$$\omega_o = \frac{1}{v_0} = \frac{t}{d_0}$$

ω_o is the time delay in the rate of expansion of the pressurized air (a.k.a. the sound wave) due to the resistance to compression of the ambient air surrounding it (a.k.a. the sound wave's medium of propagation). The interaction of these two opposing rates of motion ω_u and ω_o is what determines the speed of sound. We know from Equation 12 that the product of their reciprocals is equal to B_a/ρ_a and we know from Equation 8 that the square root of the product of their reciprocals is equal to the speed of sound.

To facilitate comparing the equation for the speed of sound with the equation for the speed of light, both velocities will be expressed in their "time constraint" form $\omega = t/d$ in the expanded equation for the speed of sound. Returning to Equation 13, it can be rewritten as follows:

$$\frac{B_a}{\rho_a} = v_s^2 = v_u\, v_o = \frac{1}{\omega_u}\frac{1}{\omega_o} \quad \text{Equation 14}$$

Substituting that for B_a/ρ_a in Equation 8 gives:

$$v_s = \sqrt{\frac{B_a}{\rho_a}} = \sqrt{v_u v_o} = \sqrt{\frac{1}{\omega_u}\frac{1}{\omega_o}} = \frac{1}{\sqrt{\omega_u \omega_o}}$$

Equation 15 which can be shortened to:

$$v_s = \frac{1}{\sqrt{\omega_u \omega_o}}$$

Does that look familiar? Recall that the equation Maxwell developed for the speed of light is:

$$c = \frac{1}{\sqrt{\epsilon_0 \mu_0}}$$

It now is time to look more closely inside the equation for the speed of light to examine the physical dynamics which determine the values of ϵ_0 and μ_0. ϵ_0 and μ_0 are not merely constants whose values were determined by experiment. Their values are determined by the strength of the innate forces of repulsion between like electric fields and between like magnetic fields. This is the same kind of innate

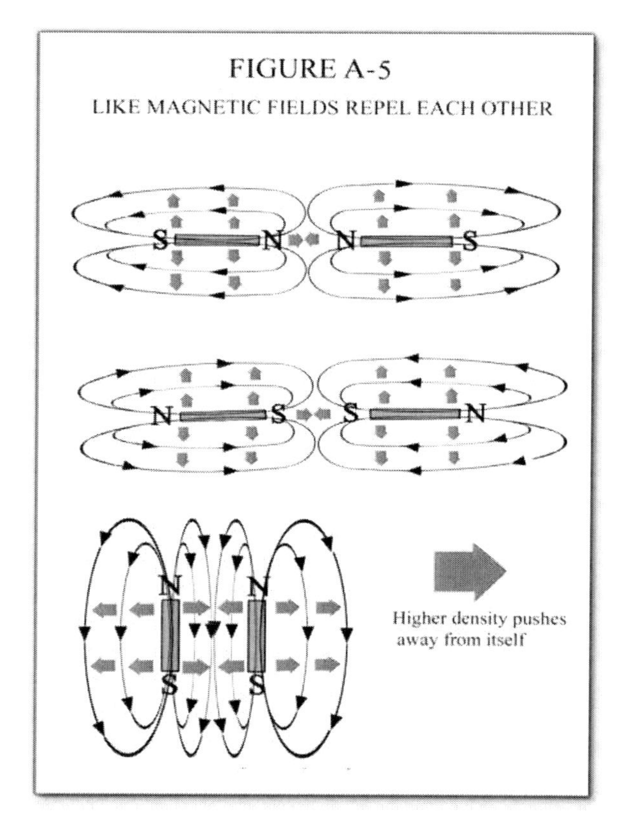

response as that of air pressure's innate compulsion to push away from itself.

4. Dissecting the Equation for the Speed Of Light

As previously shown in Figure 2, like electric fields innately repel each other. That is why electrons repel other electrons and protons repel other protons. As shown in Figure 5, the same is true of like magnetic fields. That is why a magnet resists having its north pole pushed against the north pole of another and also resists having its south pole pushed against the south pole of another. Even when the magnets are placed side by side, they resist being pushed together as long as their field lines are oriented in the same direction (i.e., are like each other).

The gray arrows show the effect of two sources of repulsion which push the magnets apart. The magnetic fields around each magnet are doing two things simultaneously. They are pushing themselves away from their greatest density inside the magnet (they are like fields to themselves) and also are pushing against the like field of the other magnet.

This innate force of repulsion between like energy fields is analogous to the innate force of repulsion between molecules of air. It causes a domain of like energy field to push away from itself, internally, (compulsion to expand) and push away from another like field externally (resistance to compression). As described for the speed of sound, the pocket of air pressure (i.e., the wave) pushes away from itself internally while it simultaneously pushes against the external pressure of the ambient air. It is the innate intensity of the force of self-repulsion which determines the speed of the wave.

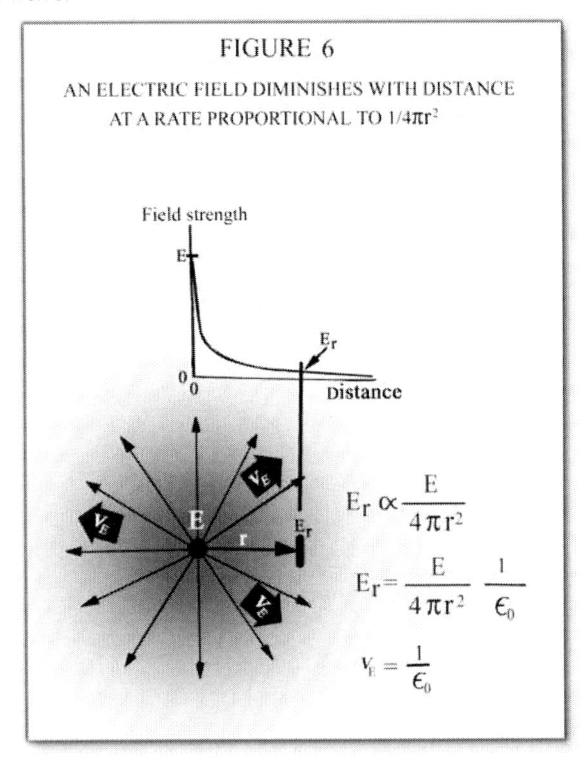

A useful concept for understanding the nature of electric and magnetic fields is to consider them as dense, self-repelling, three-dimensional clouds surrounding their source. For example, as shown in Figure 6, the field surrounding an electron can be considered as a dense cloud of energy whose intensity diminishes with distance as it pushes itself away in all directions from the electron. It's analogous to the pocket of pressurized air in empty space. As one travels outward along any field line directly away from the source, the field's density diminishes at a rate proportional to $1/4\pi r^2$.

But as shown in Coulomb's law (Equation 2) and in Maxwell's first equation, the measured field strength at a distance r from its source also is affected by whatever causes the value of ϵ_0. The benefit of picturing the field as being like a pocket of pressure is that it makes it clear that the innate force of repulsion between like fields operates both within the isolated electron's field as well as against the field of another electron. The greater field density at the source pushes the field outward. When the field is isolated and stationary, there is nothing to interfere with the field's expansion. As with the air's pressure in empty space (Figure 4) that is the condition under which the field can push itself away from itself at the greatest rate. However, there still is an experimentally determined impediment ϵ_0 to its availability at the point of measurement (i.e., at the distance r from its source). Not knowing what might cause that impediment, and thinking of the energy as moving through empty space, Coulomb attributed it to some innate characteristic of empty space. Hence, it was identified as the "permittivity of space" [xvi] However, just as the value of ω_u is determined by the rate at which air pressure can dissipate, when unopposed in empty space (Figure 4), the value of ϵ_0 determined by the innate rate at which like electric fields can dissipate when unopposed in empty space. The innate level of self-repulsion of like electric fields can be defined as the innate intensity at which adjacent parts of the field push away from each other at a given level of density (i.e., its bulk modulus). This can be expressed both in terms of velocity and in terms of its reciprocal.

$$v_E = \frac{d_E}{t} \quad \text{and its reciprocal} \quad \epsilon_0 = \frac{1}{v_E} = \frac{t}{d_E}$$

Recall that v_E and ϵ_0 are just two ways of expressing the same thing; the innate rate at which like electric fields can push away from each other. That innate level of self repulsion is what drives both the compulsion to dissipate and the resistance to being compressed. Just as with the innate rate at which air pressure can expand, we can know that the values of v_E and ϵ_0 are constants between zero and infinity since the speed of light is neither zero nor infinite. (We are not concerned with their specific values at this point. The value of ϵ_0 already has been determined by Coulomb. What we are doing here is getting familiar with what ϵ_0 is and how it works.)

As shown in Figure 7, the innate self-repulsion of like magnetic fields operates in the same manner as that of like electric fields. Thus, the innate rate at which a magnetic field can push away from itself when not being opposed by an outside force can be expressed as:

$$v_B = \frac{d_B}{t} \quad \text{and its reciprocal} \quad \mu_0 = \frac{1}{v_B} = \frac{t}{d_B}$$

Again, v_B and μ_0 both are ways to express the innate rate at which like magnetic fields can push away from each other.

Now we are at the key difference between a single medium, which possesses all five of the characteristics assumed to be present in any medium of propagation, and the two mediums for light which possesses only the fifth characteristic. (Ironically, that is the only one Michelson and Morely didn't look for.) Recall that in Equation 15, the speed at which air pressure can dissipate when unopposed v_u is being retarded by the innate resistance of the ambient air to being compressed. Thus, the speed v_o at which the ambient

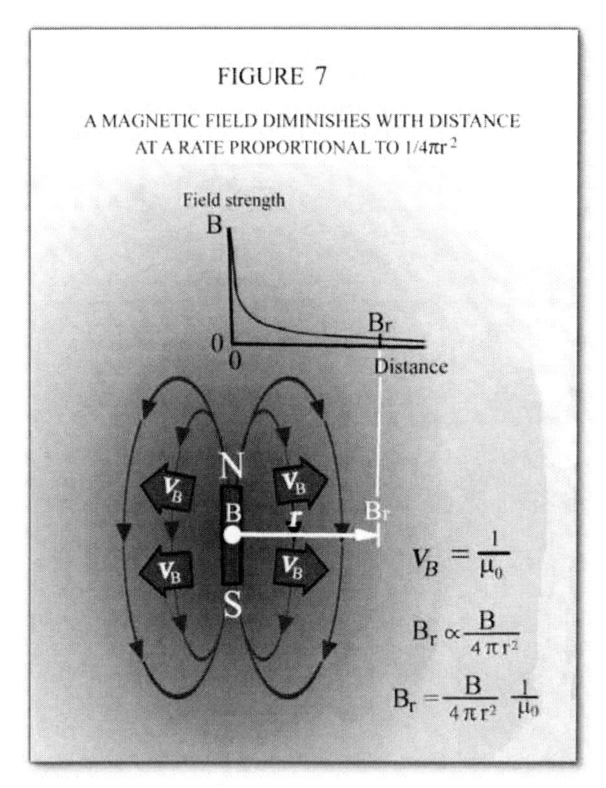

air would dissipate, from its own internal forces, is working in opposition to v_u. As shown in Equation 15, the net effect of the two forces in opposition to each other is equal to the square root of the product of their reciprocals. However, in the case of light, there are two different mediums opposing each other from opposite sides of the interaction. Whichever one is expanding (uncurling) can do so only by compressing (curling) the other.

What are the conditions of being opposed versus unopposed with respect to the movement of electric and magnetic fields? Maxwell's equations tell us that a field's movement to dissipate its intensity is unopposed when the field, itself, is not in motion. What that means is that the field and its field lines all are stationary. The field's intensity spreads out along the stationary field lines. That is akin to the pocket of pressurized air released in empty space where there is no opposition to its expansion by surrounding ambient air. But, as shown in Maxwell's third and fourth equations, when either type of field is in motion (i.e., the field itself and its field lines are in motion), it must create and curl the other type of field. As previously noted in Figure 3, increasing the curl of a looped field means reducing the radius of its field lines. That tightens the loops and increases the field's density. It is in the nature of both electric and magnetic fields to oppose having their density increased. The interaction between the two types of field consists of the compulsion of the field on one side to dissipate (uncurl) and the compulsion of the field on the other side to resist being curled. This is identical to the dynamic interaction between the pocket of compressed air and the pressure of the surrounding ambient air which both propagates the pocket of pressure (the sound wave) and determines its speed. The only difference is that with sound, the same medium (air) is both stationary and physically present on both sides of the transaction. In the case of light, the electric and magnetic fields take turns on opposite sides of the transaction as they push each other and the wave through space and travel along with it.

5. Summary And Conclusions

Having dissected the equations for the speed of sound and the speed of light, a more informed comparison now can be made between them.

Light and sound both are propagated by the innate physical behavior of their respective mediums of propagation. The behavior of the mediums of propagation, both for light waves and sound waves, is determined by their respective innate characteristics of self-repulsion. This characteristic both creates an innate compulsion for the medium to expand and an innate resistance to being compressed. It is the strength of these innate drives, which are inherent in the natures of the respective mediums, which causes the waves to propagate and which determines their respective speeds.

The propagation of light differs from that of sound and other waves in that, for light, the mediums and the wave are the same entities. (Or as the author's colleague Dr. Raymond Gallucci observed, paraphrasing Marshall McLuhan, the medium is the message.) For other waves, the medium is physically separate from the wave. Thus, for other waves, the medium is present whether the wave is propagating through it or not. The medium must be found wherever the wave is able to go. The medium is stationary and pushes the wave through it. The only thing light waves have in common with other waves is that their propagation is determined by the innate self-repelling characteristics of their respective mediums of propagation.

The reason why light can propagate through empty space is because the motions of its constituent electric and magnetic fields, once started, both create each other and push each other through space. Thus, there is no need for a stationary medium. They also don't need to permeate space because they always are right there where they are needed to push the wave. Indeed, they are the wave. The light wave and the mediums, being the same entities, travel together. Rather than being stationary, the mediums for light move at the same speed as the wave. Thus, certain characteristics of motion, such as the Doppler Effect, may not always operate in the same manner for light as for waves in stationary physical medium.

Wave	Equation	Analytical Insights	Structural Comparison
Sound	$v_s = \sqrt{\frac{B_a}{\rho_a}}$	$\frac{B_a}{\rho_a} = v_s^2 = v_u v_o$; $v_u = \frac{1}{\omega_u}$; $v_o = \frac{1}{\omega_o}$	$v_s = \sqrt{\frac{B_a}{\rho_a}} = \sqrt{v_u v_o} = \sqrt{\frac{1}{\omega_u}\frac{1}{\omega_o}} = \frac{1}{\sqrt{\omega_u \omega_o}}$
Light	$c = \frac{1}{\sqrt{\epsilon_0 \mu_0}}$	$c^2 = \frac{1}{\epsilon_0}\frac{1}{\mu_0}$; $\frac{1}{\epsilon_0} = v_E$; $\frac{1}{\mu_0} = v_B$	$c = \sqrt{v_E v_B} = \sqrt{\frac{1}{\epsilon_0}\frac{1}{\mu_0}} = \frac{1}{\sqrt{\epsilon_0 \mu_0}}$

The reason the speed of light is absolutely constant in empty space is not due to some kind of magical attribute of either space or light. It is because the characteristics of empty space are precisely the same wherever space is empty. There is nothing to interfere with or alter the innate behavior of electric and magnetic fields as they dance with each other in empty space. Wherever something interferes with the light, such as a dense cloud of cosmic dust, we say that area of space isn't empty. The same is true when light propagates itself through dense optical matter such as a lens or a prism. Contrary to popular belief, the optical matter doesn't act as a medium of propagation. The electric and magnetic fields of the charged particles in the glass randomly tug on the electric and magnetic fields in the light as it transits through the glass. The resulting meandering path averages out to the same trajectory but slows its forward propagation. That, in turn, causes the light's electromagnetic waves to refract.

In the case of sound, there are numerous climate and weather conditions that can alter the innate characteristics of ambient air. Temperature, altitude, humidity, high and low pressure areas, and wind are but a few. Similar sources of variability exist for all known waves, other than light, all of which propagate through mediums comprised of physical matter. Thus it is common to think of the speeds of waves as being innately variable. However, there is no basis for expecting the speed of light in empty space to be anything but constant. That isn't magic. It's simply the laws of physics at work in the constant, uniform environment of empty space.

---

[i] Douglas G Giancoli, *Physics*, 4th edition (Englewood Cliffs, New Jersey: Prentice Hall, 1995), 750-751

[ii] Cox, *Why does $E=MC^2$*, 29: "Most, if not all, of the scientists of the time believed that all waves, including light, must travel in some kind of medium; there must be some 'real stuff' that is doing the waving."
Giancoli, *Physics*,745: "Nineteenth-century physicists viewed the material world in terms of the laws of mechanics, so it was natural for them to assume that light, too, must travel in some medium."
Perkowitz, *Empire of Light*, 64: "The properties of this supposed medium, called the ether, could be surmised before it was found. Like any material, it could be modeled as a collection of connected atoms."

[iii] Cox, *Why Does $E=MC^2$*, 29: "It must permeate all of space, since light travels across the voids between the sun and earth and the distant stars and galaxies."
Giancoli, *Physics*, 745: "They called this transparent medium the ether and assumed it permeated all space." Perkowitz, *Empire of Light*, 65: "It seemed undeniable that only a taut, resilient ether could sustain light's speed and enormous vibrational rate. But surely such a medium would be tangible as it filled all space?"

[iv]Given that any medium of propagation was accepted as being physically separate from the wave (endnote ii) and as being present throughout the domain in which the wave can propagate (endnote iii), the medium also must be there whether or not a wave happens to be propagating through it.

[v] Cox, *Why Does $E=MC^2$*, 29: "The speed that appeared in Maxwell's equations then had a very natural interpretation as the speed of light relative to the ether. This is exactly analogous to the propagation of sound waves through air."
Giancoli, *Physics*, 745: "It was therefore assumed that the velocity of light given by Maxwell's equations must be with respect to the ether."
Perkowitz, *Empire of Light*, 64-65: "In hard, resilient stuff, like steel, the atoms are tightly linked to one another. Energy moves quickly from atom to atom. By contrast, waves move more slowly among the loosely joined atoms of a soft gelatinous substance."

[vi] Cox, *Why Does $E=MC^2$*, 29: "This is exactly analogous to the propagation of sound through air. If the air is at a fixed temperature and pressure, then sound always will travel at a constant speed, which depends only on the details of the interactions between the air molecules …"
Perkowitz, *Empire of Light*, 64-65: "The properties of this supposed medium, called the ether, could be surmised before it was found. Like any material, it could be modeled as a collection of connected atoms. In hard resilient stuff like steel, atoms are tightly linked to one another. Energy passes quickly from atom to atom, so that any waves generated in the medium oscillate rapidly and travel at high speed. By contrast, waves move more slowly among the loosely joined atoms of a soft, gelatinous substance. … It seemed undeniable that only a taut, resilient ether could sustain light's high speed and enormous vibrational rate. Somehow, the ether was firmer and more elastic than steel, yet completely penetrable. … Physicists expended great ingenuity in seeking an ether with the right characteristics, even imagining new forms of matter with the proper contradictory qualities."

[vii] The material describing Maxwell's equations presented here is excerpted from the author's book Calkins, Richard O., *Relativity Revisited* (Sammamish, Washington: A Different Perception, 2011), 14-20. For his understanding of Maxwell's equations, the author owes a great debt of gratitude to David Morgan-Mar for his excellent dissertation on Maxwell's equations in his humorous and insightful blog www.irregularwebcomic.net/1420.html, as modified December 16, 2006. Any errors in explaining the

meaning and implications of the equations are the author's own. The author also owes a debt of gratitude to Ray Gallucci, PhD, PE for identifying an error in the author's interpretation of ϵ_0 and μ_0 in *Relativity Revisited* which has been corrected here.

[viii] The subscript (EC) is added to the density symbol ρ to identify it as the density of the electric charge. This is to avoid confusion with other density parameters used later.

[ix] Giancoli, *Physics,* 465 (The subscript *r* was added to *E* in the equation for Coulomb's law for symbol consistency with Proportionality 1).

[x] Note that in Proportionality 1, we took the electric field E at the location of its point source and then made the adjustment for distance r. Coulomb's law adjusts the amount of the charge Q for the effect of distance r before measuring its field. Mathematically, these are just two different ways to get to the same result.

[xi] Giancoli, *Physics,* 628.

[xii] In other words, each of the two types of field is the medium of propagation for the other.

[xiii] The role ϵ_0 and μ_0 play in determining the speed of light has interesting implications regarding what information may be hidden in other proportionality constants. For example, does the universal gravitational constant (G) in Newton's Law of Universal Gravitation tell us something about the speed of gravitational waves?

[xiv] Giancoli, *Physics,* 314.

[xv]Ibid.

[xvi] Ibid., 461.

The Special Theory of Relativity is a House of Cards Built on a Parlor Trick and Sustained by Circular Reasoning

Richard O. Calkins
2125 Sahalee Drive East, Sammamish, WA, USA
Email: rocalkins@msn.com

A simple change in experimental design shows that the relativity principle developed by Galileo and Newton applies only to the observed motion of physical objects. Their actual motion is different in every inertial reference frame. The difference in motion is hidden by the human condition and an inadequate experimental design. The new design distinguishes between an inertial reference frame and a physical reference frame, such as a spaceship. The same experiment is conducted in one inertial reference frame and then moved intact to another. The special theory's own postulates are used to determine the changes in motion which must occur in response to the change in the spaceship's velocity. The changes required by the postulates are different from what the spaceship observer sees. The special theory treats a change in velocity differently than a difference in velocity between the same two reference frames. Applying the postulates to the change in velocity causes the first postulate of relativity to refute the premise of equal merit. The first postulate requires changes in response to the change in motion which the spaceship observer cannot detect. His observation is invalid. The second postulate then refutes the first postulate. The second postulate imposes a constant speed on light which alters the time interval between emission and impact when the experiment is moved to the second reference frame. This difference cannot be corrected by time dilation without making the experiment using a physical object produce a different time interval in the second reference frame. The change in experimental design also reveals a previously unrecognized circularity between an observer's own state of motion and his perception of motion. This circularity has undesirable consequences. It blinds an observer to changes in motion which are caused by a change in his own state of motion. It also causes the observer to see changes in motion for objects whose motion has not changed. And it reveals that each observer in a different inertial reference frame has his own definition of motion which is unique to him and entirely subjective. Such observations are virtually worthless for scientific purposes. Lastly, the paper discloses how unquestioning belief in the theory's own postulates and premises has controlled the design of experiments and the interpretation of data to prove the theory's validity. The role of circular reasoning helps explain how an internally contradictory theory has experienced more than a century of empirical validation. This paper demonstrates its case simply by making two changes in experimental design. There are no flights of fancy into new theoretical realms to be found in this paper. The results of the experiments are based entirely on the special theory's own postulates. The foundational beliefs of the special theory itself determined the outcome of this paper.

1. Introduction

Einstein's special theory of relativity is based on a combination of Galileo's and Newton's relativity principle and Maxwell's equations.

The relativity principle developed by Galileo and Newton addressed only the motion of physical objects. It is based on the fact that any experiment involving the motion of physical objects will produce the same result in every inertial reference frame.The laws that govern the motion of physical objects are called the laws of mechanicsi and their consistency of form in all inertial reference frames is called the relativity principle:

"The basic laws of physics are the same
in all inertial reference frames." [ii]

Einstein's first postulate of relativity expands the relativity principle into a universal principle of motion. It is declared to apply not only to electricity and magnetism but to all phenomena involving motion, including even phenomena not yet discovered.iii

Einstein's second postulate is based on the definite constant speed of light as predicted by Maxwell's equations.iv

Because experiments in all inertial reference frames are believed to produce the same result, the special theory's premise of equal merit treats all observations made in inertial reference frames as being equally valid.v Considering that the theory's predictions are produced by the process of reconciling conflicting observations of the same experimental events, the possibility for observation error should warrant consideration. Instead, the premise of equal merit simply defines observation error out of existence. Barring a malfunction of the experimental apparatus, observations made in all inertial reference frames are declared on their face to be equally valid.

Based on the belief that motion can exist only relative to the reference frame of an observervi and due to its focus on inertial motion, the special theory uses inertial reference frames as the only venue for conducting experiments and making observations. That experimental design omits information from non-inertial reference frames. Einstein deferred the issues of non-inertial motion to the general theory. However, as this paper shows, some of the information which is required to understand the nature of inertial motion can be found only in a non-inertial reference frame.

This paper demonstrates its case simply by making two changes in experimental design. The results of the experiments are determined entirely by the special theory's own postulates. There are no flights of fancy into new theoretical realms to be found in this paper.

The first change in the experimental design is to replace the customary light bulb with a tightly focused optical laser. This reduces the ambiguity sometimes found in the special theory regarding the direction in which the observed light is propagating. The second change is to provide experimental data which is not available to observers in inertial reference frames. To do so, the experimental design takes advantage of the difference between a physical reference frame and an inertial reference frame. The two are different in kind and should not be lumped together or confused with each other.

A physical reference frame, such as a spaceship, can change its own velocity from that of one inertial reference frame to that of another. This allows an observer to make the same experiment using the same equipment in one reference frame and then change its own velocity to make the same precise experiment in a different inertial reference frame. This breaks the quarantine between the different reference frames by establishing a direct link between them. Each reference frame's state of inertial motion is identical to that of the other except for the change in the spaceship's velocity.

As shown in Section 3, this new link discloses two important new insights. First, it discloses that the special theory interprets a change in velocity differently than it interprets an equal difference in velocity between the same two inertial reference frames. Second, it discloses a previously unrecognized circular dependence between an observer's state of motion and both his definition of motion and his sensory perception of motion. This circularity leads to two kinds of observation error. One is that it renders the spaceship observer selectively blind to the effects of a change in his physical reference frame's velocity. The other is that it causes the spaceship observer to perceive changes in motion of objects whose motion has not changed.

As mentioned above, The Galileo/Newton relativity principle is the prime foundation of the special theory. It is the basis for the first postulate. The fact that experiments involving the motion of physical objects will provide the same observed results in all inertial reference frames has been taken to mean that the actual motions of the objects are the same in all inertial reference frames. However, Section 4, by applying the new design to a physical object, discloses that the actual motion of the physical object will be different in every inertial reference frame. Only the observations made in different inertial reference frames will be the same. This unveils the relativity principle as a clever parlor trick the human condition has played on us mere humans. The sameness of observed results in every inertial reference frame isn't a fundamental principle of motion; it's due to the circular relationship between an observer's state of motion and his perception of what constitutes motion. As often happens in mundane existence, perception is not reality.

Section 5 then applies the same experimental design to an experiment involving the propagation of light. What that discloses is that light can't even do the parlor trick. If an experiment is conducted in one reference frame and then moved to another, it will provide the same observed trajectory but it cannot produce the same elapsed time between emission at the source and arrival at the target. Unlike a physical object, light is subject to the second postulate. It has a definite, constant speed. A change in its source's velocity will change the length of the light's trajectory (which the spaceship observer will not detect) but will produce a different interval of time for the light to complete the trip. The elapsed time is an observed result. Because an observed result is changed when the velocity of the source is changed, even the "same observed result" requirement of the relativity principle cannot be met. As a result, light cannot even do the parlor trick. As shown in Section 5, this problem cannot be resolved by invoking the theory's prediction of time dilation.

Section 6 exposes an innate inconsistency between how the special theory treats the motion of some locations in an observed reference frame versus how it treats the motion of other locations in the same reference frame. This inconsistency is innate to observations made from inertial reference frames. It is an unrecognized side effect of that experimental design.

Section 7 then provides an example of how a change in an observer's state of inertial motion will cause him to perceive illusionary changes in the motion of objects whose motion has not changed.

Finally, Section 8 discloses the role of circular reasoning in the design and conduct of experiments used to validate the special theory for more than a century. Circular reasoning occurs when an unquestioning subconscious acceptance of the beliefs being examined is allowed to influence the experimental design and the interpretation of empirical data used to examine them. The presence of circular reasoning is clearly disclosed both in the experimental design and in the interpretation of empirical data for experiments used to validate the special theory.

2. An Overview of the Special Theory of Relativity

Einstein's theories of relativity are the very foundation of modern physics.vii His special theory of relativity is the basis of our understanding of such fundamental phenomena as motion, time, space and mass. His general theory of relativity provides our understanding of gravity. Professional physicists consider Einstein's theories of relativity as having been consistently sustained by experimental validation over the past century. They are considered to be the gold standard of generally accepted theory.

Prior to Einstein's theories of relativity both Galileo and Newton had determined that the laws of physics that govern the motion of physical objects take the same form in all inertial reference frames. In other words, if you are in a reference frame that is either stationary or in motion at a constant speed in a straight line, any experiment you make involving the movement of physical objects will produce the same result as it would if performed in any other inertial reference frame. You will feel that your reference frame is stationary and any experiment you conduct with moving physical bodies will produce the same result as if your reference frame were stationary. More specifically, it means that the mathematical equations that describe the result will take the same form regardless of your reference frame's state of inertial motion. The laws that govern the motion of physical bodies are called the laws of mechanicsviii and their consistency of form in all inertial reference frames is called the relativity principle:

"The basic laws of physics are the same in all inertial reference frames."[ix]

Note that the relativity principle is based on the laws of mechanics and the work of Galileo and Newton. It had nothing to do with the propagation of light until Einstein concluded that it is a basic law of physics which should apply to all forms of motion.

In 1864, the renowned physicist James Clerk Maxwell produced his insightful and comprehensive theory of electromagnetism.[x] His theory showed that the movements of electrically charged particles create electromagnetic waves which propagate through empty space at a precise, constant speed of 299,792,458 m/s.[xi] Because that was exactly the same as the previously determined speed of light, Maxwell realized that light could be considered to be an electromagnetic wave and that it always will propagate through empty space at that precise speed.[xii] Maxwell also predicted the existence of other kinds of electromagnetic waves which were not visible to the human eye. We now know of many kinds, such as radio waves, microwaves, infrared, ultraviolet, X-rays and gamma rays.[xiii] All electromagnetic waves propagate through empty space at the same speed. They differ only in their respective frequencies (or wavelengths) which alter the manner in which they interact with matter. The problem raised by Maxwell's equations was that they did not take the same form in all inertial reference frames.[xiv]

If light was a wave, as shown by Maxwell's equations, that wouldn't necessarily be a problem. Generally accepted theory would require it to have a stationary, physical medium of propagation, just as does every other kind of wave. The particles in a physical medium propagate waves by moving in accordance with the laws of mechanics,which are the basis of the Galileo/Newton relativity principle. Thus, it was generally accepted that there must exist some kind of undetectable medium for the propagation of light that permeated all of open space. It was called the ether.[xv] The reference frame for which Maxwell's equations took their simplest form was taken to be the one which was at rest in the ether.[xvi] The additional terms required for other reference frames were to adjust for their movement relative to the ether. This would account for the differences in the form of Maxwell's equations and would resolve the apparent conflict with the Galileo/Newton relativity principle.

To resolve the matter, an endeavor began in the late 1800s to detect the existence of the ether and to determine the speed at which our own reference frame, the Earth, is moving through it. The most famous and most conclusive experiment was conducted by A. A. Michelson and E.W. Morley in the 1880s.[xvii] Unfortunately, what their experiment proved was that there was no ether. There was no explanation for how light could propagate through empty space or why Maxwell's equations took a different form in different inertial reference frames. This was one of the great puzzles of physics at the beginning of the 20th century.[xviii]

Einstein solved the puzzle using his famous thought experiments.[xix] Einstein concluded that the inconsistencies in Maxwell's equations resulted from the assumption that an absolute space exists.[xx] In his famous 1905 paper, Einstein proposed eliminating both the ether and the corresponding belief in the existence of a reference frame at rest.[xxi] Based on this proposal, Einstein stated his two postulates of relativity. His first postulate declares:

"The laws of physics have the same form in all inertial reference frames."[xxii]

As discussed above, the first postulate extends Galileo's and Newton's relativity principle, which was based on the laws of mechanics, to govern the motion of all physical phenomena, including not only electricity and magnetism but even phenomena not yet discovered.xxiii Any experiment involving motion will produce the same result in all inertial reference frames regardless of what is in motion and what caused it to be in motion. Motion exists only relative to a defined frame of reference and inertial motion exists only relative to an inertial reference frame.

Einstein's second postulate of relativity, which is based on Maxwell's equations, specifically addresses the propagation of electromagnetic waves:

"Light propagates through empty space with a definite speed *c* independent of the speed of the source or observer."[xxiv]

These two postulates are the conceptual foundation of Einstein's special theory of relativity. Related foundational premises which underlie and add specificity to the postulates include:

- The Galileo/Newton relativity principle applies to objects which are subject to the laws of mechanics. Thus, any phenomenon which meets the Galileo/Newton relativity principle must behave as if it possesses mass and experiences momentum. By extending the relativity principle to the propagation of light, Einstein's first postulate requires light to respond to a change in the velocity of its source in the same manner as if it were a physical object being launched from the source.[xxv]
- There is no such thing as an absolute state of rest.[xxvi] This is the single, most important hypothesis underlying the theory. It is the basis for the first postulate.[xxvii]
- All inertial motion is relative and can be identified only in terms of motion relative to a specified reference frame. This follows from the nonexistence of a unique reference frame which is at absolute rest.[xxviii]
- Since observations made in all inertial reference frames produce the same result, observations made from all inertial reference frames have equal merit (i.e., are equally valid).[xxix] For brevity, this is referred to in this paper as the premise of equal merit.
- Since observations made in all inertial reference frames produce the same result, an observer in any inertial reference frame who measures the speed of light will obtain the same number c = 299,792.5 km/s.[xxx] For brevity, this is referred to in this paper as the premise of will measure c.

These are the beliefs which must be valid in order for the special theory to be valid.

3. An Overview of the Galileo/Newton Relativity Principle

The relativity principle produced by Galileo and Newton states that experiments involving the motion of physical objects will produce the same result in all inertial reference frames. For example, Figure 1 shows two observers in different inertial reference frames conducting the same experiment. It consists of firing a tiny steel ball from a tiny cannon. The cannon is on the floor of a spaceship and is pointed vertically at a target on the ceiling. The observer in reference frame B will observe the same result as the observer in reference frame A regardless of the difference in their relative inertial motion. The result will be the same in every inertial reference frame.

However, as shown in Figure 2, if each observer observes the experiment being conducted in the other observer's reference frame, the result each observer sees will be different from what happens in his own reference frame. For example, suppose that the observers in the two reference frames are moving away from each other at velocity v_d. (v_d is the difference between their respective, but unknown, inertial velocities.) Each observer will feel as if he is stationary and will see the cannon in the other reference frame moving at v_d relative to him. According to the special theory, the difference in the steel ball's trajectory results from the cannon's motion relative to its observer. The

FIGURE 1

ANY EXPERIMENT INVOLVING THE MOTION OF PHYSICAL OBJECTS WILL PRODUCE THE SAME RESULTS IN ALL INERTIAL REFERENCE FRAMES

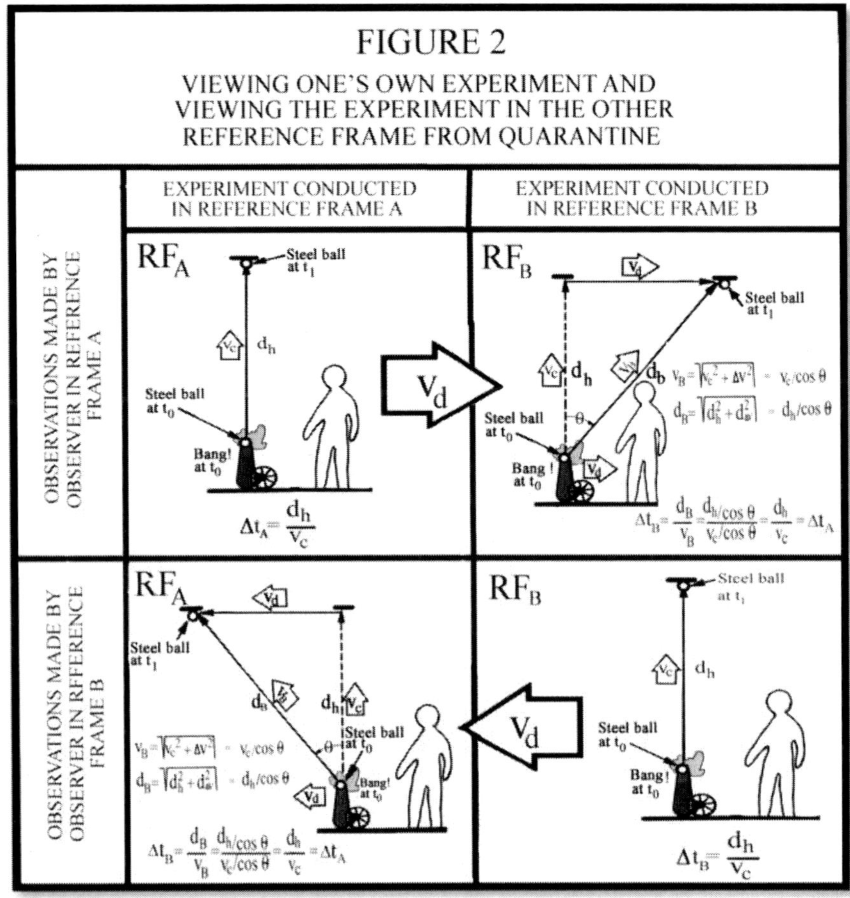

premise of equal merit holds that all four observations shown in Figure 2 are equally valid. The difference in the steel ball's trajectory when viewed from another reference frame is caused by the difference in its momentum which is caused by the difference in the cannon's velocity relative to the observer. The steel ball in the experiment conducted in reference frame B (lower right hand box) will travel on a diagonal trajectory in observer A's reference frame because the cannon which fires it is in motion in observer A's reference frame (upper right hand box). The steel ball in the experiment conducted in reference frame A (upper left hand box) will travel on a diagonal trajectory in observer B's reference frame because the cannon which fires it is in motion in observer B's reference frame (lower left hand box). Each observer sees the steel ball fired by the cannon in his own reference frame travel vertically because that cannon is not in motion in his reference frame.

4. A Peek Inside the Galileo/ Newton Relativity Principle

The special theory is based exclusively on experiments made in inertial reference frames with the results being observed by observers in the same or other inertial reference frames. All an observer in an inertial reference frame can know about the motion of another inertial reference frame is by observing its motion relative to him. Because each observer sees himself as being stationary (v=0) and the other as being in motion (v>0), each observer interprets the other reference frame as moving faster than his. Thus, the special theory's predictions of time dilation, space compression, and increasing mass are attributed to an increase in velocity relative to the observer. However, if two inertial reference frames are in different states of inertial motion, it is axiomatic that one of them is moving faster than the other and the other is moving more slowly. As a simple matter of mathematics, it is no more possible for each observer to move faster than the other than it is for all of the children in Lake Wobegone to be more intelligent than their average. However, since neither observer can know which inertial reference frame he occupies, there is no way for them to recognize that mathematical fact. But that mathematical fact means that the theory's predictions of time dilation, increasing mass and length shortening must occur whether the "other" inertial reference frame is at a higher or a lower inertial velocity than that of the observer's reference frame. According to the special theory and simple mathematics, time dilation, reduction in length and increasing mass must place limits not only on how fast one can go but on how much he can slow down.

Experiments and observations made exclusively in inertial reference frames are innately insufficient to capture a full empirical description of the phenomena being observed. They address only differences in motion between different inertial reference frames. They do not consider the effects of the change in motion which must take place in order to move from one inertial reference frame to another. The only way to compare what the postulates require in response to a change in motion with the observed effects of a difference in motion between the same two reference frames is to break the quarantine between them. And the way to break the quarantine is to take advantage of the difference between a physical reference frame and an inertial reference frame.

Figure 3 does the same experiment shown in Figures 1 and 2 but uses a spaceship to conduct the experiment in reference frame A before moving it to reference frame B. It then conducts the same experiment in reference frame B.

Box A in Figure 3 shows the results of the experiment when conducted in reference frame A. Note that the experiment is identical to the one shown on the left side of Figure 1. In Box A of Figure 3, an observer has been added to reference frame A. He will remain in reference frame A when the spaceship moves to reference frame B.

Box B of Figure 3 shows something that observations made exclusively from inertial reference frames cannot detect. In order to get from reference frame A to reference frame B, it is necessary for the spaceship to change its velocity. In so doing, some things happen which are exclusive to the spaceship and the spaceship observer. The spaceship observer feels acceleration as the spaceship is changing its velocity. No one else does. His accelerometer tells him that he is accelerating at a constant rate a_s. His timer tells him the interval of time Δt_s which is required to become stationary relative to the objects in reference frame B. The product of a_s and Δt_s tells him the magnitude of his change in horizontal velocity Δv. Not surprisingly, Δv is the same as the difference in velocity v_d between the two reference frames. However, the fact that it is a change instead of a difference is significant. The change in the spaceship's velocity can have an effect only on the spaceship and on that which resides inside of it. Nothing outside of the spaceship has changed its state of motion.

When the spaceship arrives in reference frame B, the spaceship observer shuts down its rocket. The change in the momenta of the spaceship, laser, target and observer is what maintains their new velocity after the rocket has been extinguished (laws of mechanics). The same holds true for the steel ball after it leaves the cannon and no longer is in contact with it (laws of mechanics).

As shown in Box C, when the spaceship observer conducts the experiment in reference frame B, the observer who remained in reference frame A sees exactly what the observer in reference frame A did in Figure 2 (upper right hand box). And as shown in Box D, the spaceship observer, who now is in reference frame B, sees the same result the observer in reference frame B did in Figure 1.

However, Box B shows something entirely new. It is something which clearly must happen in order to move the spaceship from reference frame A to reference frame B. But it happens in a non-inertial reference frame. Empirical information is not even collected in non-inertial reference frames, let alone being reported and included in the interpretation of the data which are reported. *But what is shown in Box B is crucial information about why the change in trajectory occurs and in which of the two reference frames the change must happen.*

When the velocity of a physical object is changed, the laws of mechanics require it's momentum to change by $\Delta P = m\,\Delta v$, where m is the object's mass and Δv is the change in its velocity.xxxi What happens in Box B is what changes the horizontal velocity of the spaceship and everything in it. It also is what changes the momenta of the spaceship and everything in it. Unlike the interpretation of the difference in velocity which was observed in Figure 2, the change in the steel ball's trajectory in Figure 3 is caused by the change in velocity from moving to reference frame B; The change in trajectory does not happen in reference frame A, where it is observed. It occurs in reference frame B, where the change in

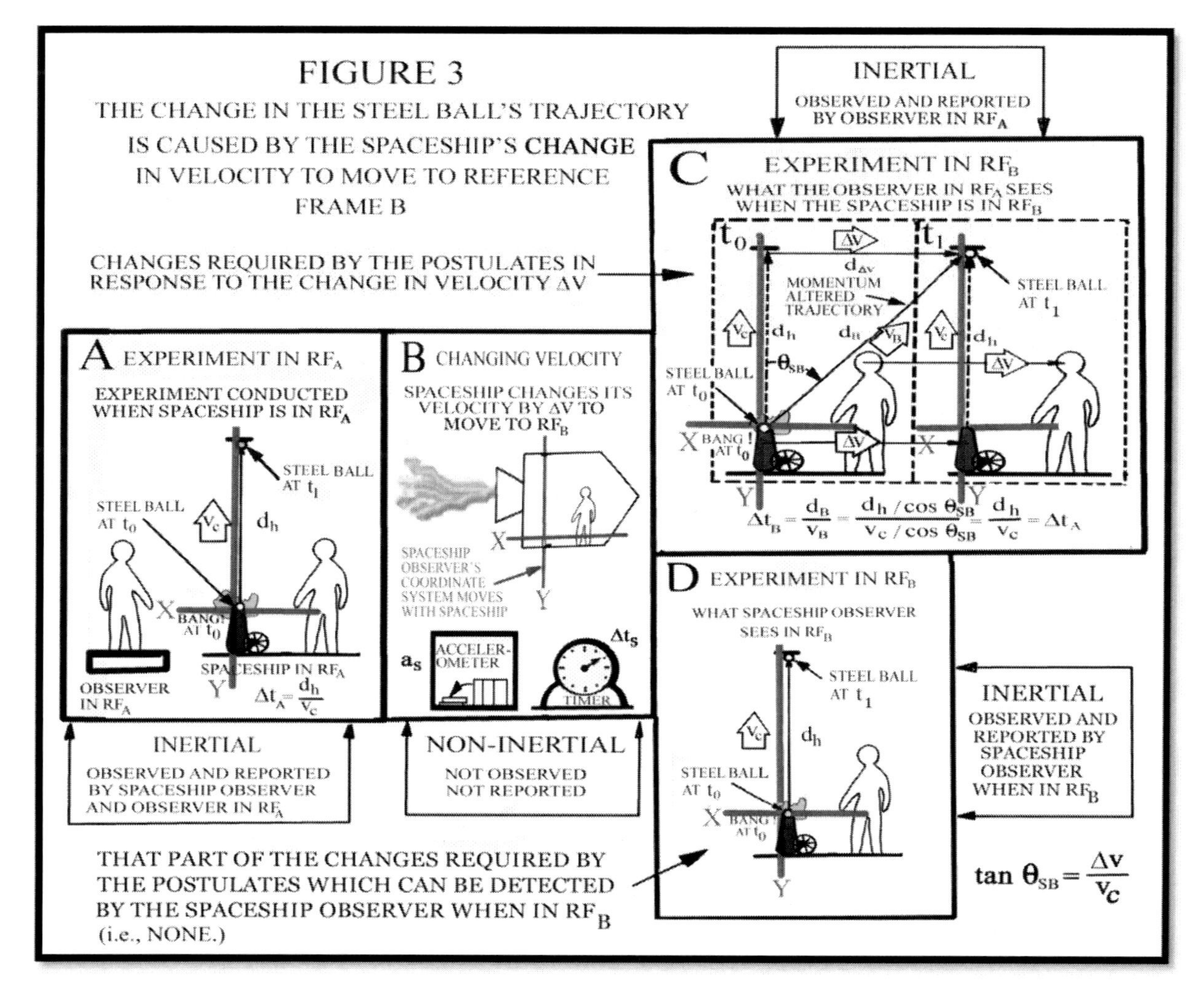

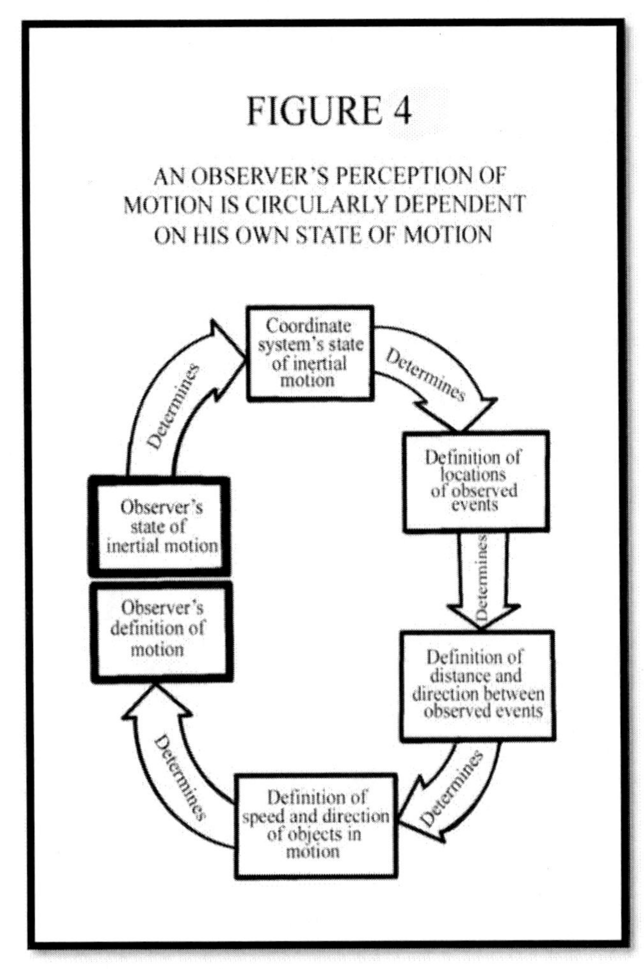

momentum is in effect.

Note that in Figure 3 the observer who remained behind in reference frame A is the one who sees the change in the steel ball's trajectory. Also note that the spaceship observer does not detect any change after he and the spaceship have moved to reference frame B, despite the fact that that is where the laws of mechanics (i.e., the source of the relativity principle and the first postulate's foundation) require the change to happen.

The questions this raises are:

- "Why is it the observer who remained in reference frame A who correctly observes the change in the steel ball's trajectory which is caused by the spaceship's move to reference frame B?" and
- "Why can't the spaceship observer detect what the laws of mechanics say must happen in his own physical reference frame after he moves it from inertial reference frame A to inertial reference frame B?"

The answers to those questions can be found only by observing and recording what happens in the non-inertial reference frame shown in Box B. When the spaceship observer moves his physical reference frame from inertial reference frame A to inertial reference frame B, he takes his human senses and the coordinate system he uses to recognize and measure motion along with him. As shown in Figure 4, this imposes a circular dependence between his own state of motion and his definition of what constitutes the presence, direction and amount of motion.

The reason why the observer who remained behind in reference frame A can see the change in trajectory is because he and his coordinate system did not change. The reason why the spaceship observer cannot see the change in trajectory is because the tools he uses to sense and define motion have made identically the same change in velocity as the spaceship, cannon, target and steel ball. When the physical reference frame in which the laboratory is located changes its own motion from one inertial reference frame to another, there is no means by which an observer in the laboratory can detect the change in motion after it is in effect (i.e., from the vantage point of the new inertial reference frame).

Note that the same circularity described in Figure 4, which masks the change in the steel ball's trajectory, also masks the change in its horizontal momentum. Because Δv changes the horizontal velocity of both the spaceship and the spaceship observer, as well as that of the steel ball, the spaceship observer also cannot detect the change in the steel ball's horizontal momentum. If he held his hand out to contact the ball, he would feel only the vertical momentum imposed by the powder charge in the cannon. Again, that doesn't mean that the change in horizontal momentum doesn't happen. The laws of mechanics require it to happen. The spaceship observer can't detect it simply because of his own change in horizontal velocity hides the effect of the steel ball's change in horizontal velocity and the corresponding change in its horizontal momentum.

What is shown in Figures 3 and 4 unveils the fact that the Galileo/Newton relativity principle is based on a subtle parlor trick the human condition has been playing on us mere humans.

- For the same experiment involving the motion of physical objects, the *actual* result will be different in every inertial reference frame. Only the *observed* result remains the same. The difference between what happens and what is observed is caused by the circularity between an observer's state of motion and his perception of motion (an unrecognized flaw in experimental design).
- The human body experiences the same physical sensations in every inertial reference frame. It cannot tell one from another (human condition).
- When in an inertial state of motion, the five senses of the human body cannot detect any motion other than relative to the human who harbors them (human condition).
- When an observer changes from one inertial reference frame to another, he takes his human senses and the coordinate system he uses to identify locations and define motion along with him (an innate flaw in experimental design coupled with the limitations of the human condition).

- Each observer's definition of motion in a given inertial reference frame is entirely subjective and is determined by his own state of motion. Such observations are innately unreliable for scientific analysis (a result of both the human condition and experimental design).
- The method for identifying motion is restricted to experiments conducted in and observations made from inertial reference frames. Such observations can detect only *differences* in inertial motion. Thus, they fail to discern that the special theory treats a *difference* in the source's velocity between any two reference frames differently from how it treats an equal *change* in the source's velocity between the same two inertial reference frames. Based on the premise of equal merit, a *difference* in momentum, which is caused by a *difference* in the source's velocity, occurs in the reference frame in which the *difference is observed.* The laws of mechanics require the *change* in an object's trajectory, due to a *change* in its source's velocity, to occur in the reference frame in which the *changed velocity is in effect.* This is the reverse of how the special theory treats a *difference* in inertial velocity (an unanticipated effect of an inadequate experimental design).

Note that the observer in any inertial reference frame has arrived there by changing his velocity from another reference frame. Thus, the circularity problem exposed in Figures 3 and 4 applies to observations made from virtually any inertial reference frame.

With apologies to Shakespeare and Hamlet, one might say that there is more motion in inertial reference frames than is dreamt of in our philosophy.

5. Why Light Cannot Even do the Parlor Trick (Light Has a Constant Speed)

This experiment is identical to the one shown in Section 4, which used vertically oriented cannon to fire a steel ball, except that a vertically oriented laser is used to fire a quick burst of light. The laser is substituted for the light bulb customarily used in textbook illustrations to reduce ambiguity regarding the vector direction in which the light being observed is propagating. As shown in Figure 5, absent interference from other forces, the light a laser emits will propagate on a specific vector trajectory. That trajectory is determined by the laser's physical architecture. That is why one can refer to the "direction" in which a laser is "pointed". However, by knowing about a laser's architectural trajectory, one can know that if light responds to momentum, a change in the laser's horizontal velocity will change the angle between its architectural trajectory and the trajectory traveled by the light it emits. That effect is due the change in the light's momentum, which is interference from another source.

Just as in the steel ball experiment, the experiment using the laser is done first when the spaceship is stationary in reference frame A. As shown in Box A of Figure 6, the laser is oriented to emit a quick burst of light vertically to strike a target on the ceiling of the spaceship. Δt_A, the time required for the light to complete the trip in reference frame A, is equal to the distance d_h between the laser and the target divided by the speed of light c. Since the premise of equal merit requires that what is shown in Box A is a valid description of reality, it is assumed that the light burst shown in Box A actually travels on the laser's architectural trajectory.

As shown in Box B, the spaceship observer uses the spaceship's rocket to change its velocity by Δv to move from

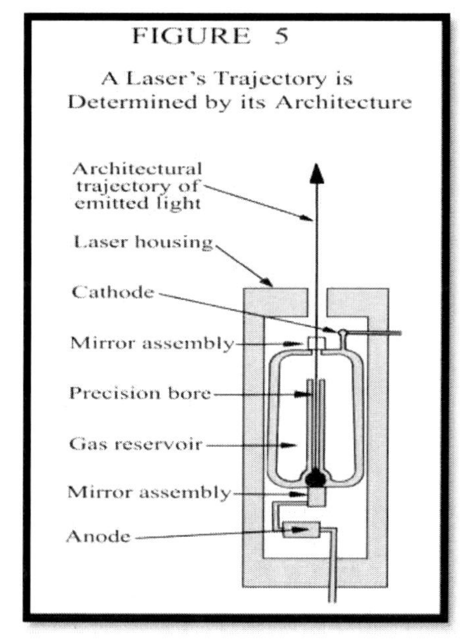

reference frame A to reference frame B. As before, he leaves reference frame A's observer behind to keep an eye on things from reference frame A's point of view. Also, as before, the spaceship observer's sensation of acceleration is unique to him. He also is unique in taking his human senses and the coordinate system he uses to define motion along with him. And he is the one who is best situated to observe and record the rate of acceleration and the time interval during which it was applied. This is information which is neither collected nor reported in experiments customarily used to examine inertial motion.

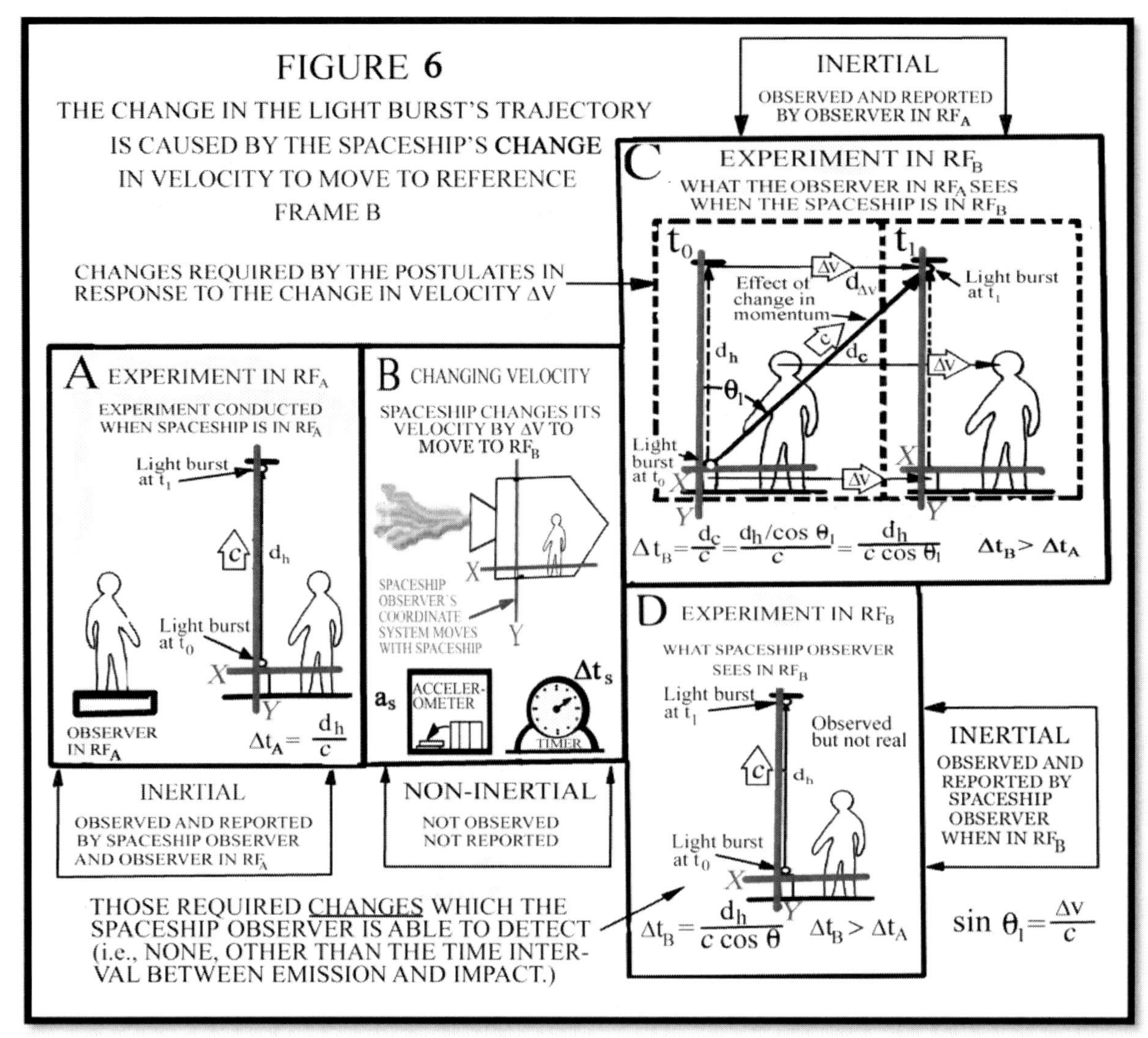

As shown in Box C, what happened in Box B has changed the horizontal velocity of the spaceship and of everything inside it. That changes the laser's momentum and, according to the first postulate, also changes the light burst's momentum. Thus, as the light departs from the laser, it no longer will travel on the laser's architectural trajectory; the light now must travel on a new trajectory in response to the change in its momentum. The new trajectory will be a combination of the change in the light's horizontal velocity Δv and the speed of light c. However, unlike the steel ball, the light's velocity will not be the vector sum of its previous vertical velocity and the momentum-induced change in the laser's horizontal velocity. According to the second postulate, light must continue to propagate on its new trajectory at its definite constant speed c. Thus, the angle of the light's divergence from the laser's architectural trajectory θ_l, will have a sine equal to the change in horizontal velocity Δv divided by the speed of light c. Recall from Figure 3 that the angle of divergence θ_{SB} shown in Box C, relative to the direction in which the cannon was pointed, had a tangent which was equal to Δv divided by v_c. Δv was the change in the cannon's horizontal velocity and v_c was the vertical velocity imposed by the gunpowder in the vertically pointed cannon.

The result of the light's restriction to a definite constant speed is that, unlike the steel ball, the length of its trajectory is increased by the effect of the change in its momentum but its speed is not. The interval of time it takes to complete its trip from laser to target Δt_B in reference frame B must increase.

$$\Delta t_B = \frac{d_c}{c} = \frac{d_h/\cos\theta_l}{c} = \frac{d_h}{c\cos\theta_l}$$

$\Delta t_B > \Delta t_A$

As shown in Box D, the spaceship observer remains blissfully unaware of the change in the light burst's trajectory. This is because, as shown in Box B, The spaceship observer and his coordinate system experience identically the same change in horizontal velocity as the spaceship, laser and target. The spaceship observer feels the same as when he was in reference frame A and his coordinate system shows no changes in horizontal position from what they were in reference frame A. He still is in thrall to the illusion cast by the human condition's parlor trick. However, if he has a really good mechanical timer, he will notice that there is something wrong with the experiment's elapsed time. It is longer than it was in reference frame A. That is an empirical result which is not the same as in the previous inertial reference frame. Thus, the experiment involving light does not satisfy the relativity principle. The constant speed imposed by the second postulate refutes the first postulate. Also, the failure to satisfy the relativity principle invalidates the basis for the premise of equal merit. Experiments involving the propagation of light do not produce even the same observation in different inertial reference frames, let alone producing the same actual motion. With the loss of the relativity principle and the premise of equal merit, the special theory collapses like the proverbial house of cards.

This is the point in the discussion when the true believers in relativity will explain time dilation to this poor, uninformed amateur analyst. However, both the steel ball experiment and the light experiment have made the same change in velocity Δv from the same reference frame A to the same reference frame B. Indeed, the spaceship observer made both the steel ball experiment and the light experiment in reference frame A before changing the spaceship's inertial velocity to that of reference frame B. He then did both experiments again. As shown in Figure 3, the steel ball experiment produces the same apparent trajectory to the spaceship observer in reference frame B (Box D) as it did in reference frame A (Box A). Assuming the spaceship observer has a really good mechanical timer, the steel ball experiment also will produce the same time interval in reference frame B as it did in reference frame A. For the steel ball experiment, the illusion created by the human condition's parlor trick is fully intact. However, the light experiment fails the time interval test. It creates the same illusion for its trajectory but comes up with a different elapsed time (Box C of Figure 6). Now, here's the rub for the true believers; for the same inertial reference frame, there can be only one rate at which time passes. If the clock is adjusted to produce the correct answer for the light experiment, the steel ball experiment will fail the "same observed result" test for its elapsed time. There simply is no way one can play with the clock to produce the right answer for both experiments. It's a simple matter of mathematics. The propagation of light has a definite constant speed. Steel balls fired from a tiny cannon do not.

One reasonably might wonder why I am fixated on having a really good mechanical timer. The reason is that mechanical timers are immune to both changes in inertial velocity and differences in inertial velocity. Their timing may be affected during acceleration, if it is extreme enough, but not when in inertial motion. However, as shown in Box C of Figure 6, a light clock slows down as its inertial velocity is increased. If the instant of emission is "tick" and the instant of impact is "tock"' the distance light travels between the "tick" and "tock" events will increase but the speed at which it travels between those two events will not increase. The greater the increase in the light clock's inertial velocity, the more inaccurate it will become. Also, it is mathematically clear in Box C of Figure 3 that the only way for the steel ball to give the correct time interval is for time to pass at the same rate in reference frame B as it does in reference frame A.

I'm sorry, true believers. Time dilation is dead. When faced with a change in the physical laboratory's inertial velocity, the observer in the laboratory cannot detect the changes in the trajectory of either a steel ball or of a burst of light which are mandated by the first postulate. Even if the first postulate is correct, the premise of equal merit is invalid. In the case of light, the second postulate's constant speed then produces a different elapsed time in every inertial reference frame, thereby refuting the first postulate's claim that the relativity principle applies to the propagation of light. And there is no way to adjust the clock in the second inertial reference frame to provide the same elapsed time for both the experiment using the steel ball (matter) and the experiment using the burst of light (energy). Time dilation has died because it is unable to resuscitate itself.

This would seem to be a good place to conclude and retire. But, like the guy who gives the spiel in the commercials, I feel compelled to cry "Wait! There's more!" It still is worth understanding why observations made exclusively from inertial reference frames and interpreted under the influence of the premise of equal merit can be even worse than worthless. They not only fail to produce all of the information needed to understand the phenomena being observed, they also can be downright misleading, as shown in the next two sections.

6. Observations Made From Inertial Reference Frames Treat Relative Motion Inconsistently

The concept required to appreciate this problem is to recognize that an inertial reference frame is simply a state of motion. Its coordinate system spreads out in three-dimensional space to the very ends of the universe. Everything in the universe that is stationary relative to that

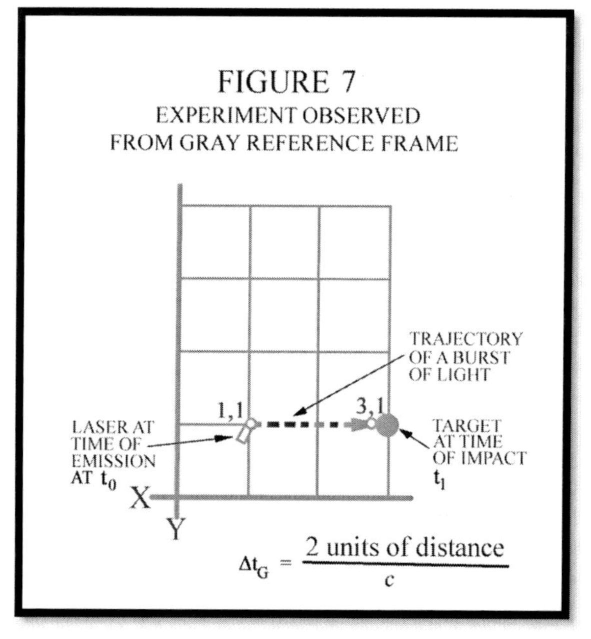

FIGURE 7
EXPERIMENT OBSERVED FROM GRAY REFERENCE FRAME

coordinate system is in that same inertial reference frame.

Every other reference frame also has its own coordinate system which extends to the very ends of the universe. If two inertial reference frames are in a given state of motion relative to each other, the entire universe-wide coordinate system of each reference frame is in the same state of motion relative to the entire universe-wide coordinate system of the other.

This experiment shows the observations made in one inertial reference frame, shown in gray, of an experiment made in another inertial reference frame, shown in black. The experiment consists of a laser situated at one location in the black reference frame's coordinate system which emits a burst of light. The burst of light subsequently strikes a target situated at another location in the black reference frame's coordinate system.

Figure 7 shows the locations of the two events as recorded by the observer in the gray reference frame. He sees a laser located at coordinates 1, 1 in his reference frame emit a burst of light at time t_0. At time t_1, he sees the burst of light strike a target located at coordinates 3, 1 in his reference frame. During that interval of time, the light was seen to travel horizontally a length of two units of distance.

Figure 8 shows where the laser and target are located in the black coordinate system. In the black reference frame, the light travels diagonally a distance of approximately 2.236 units of distance.

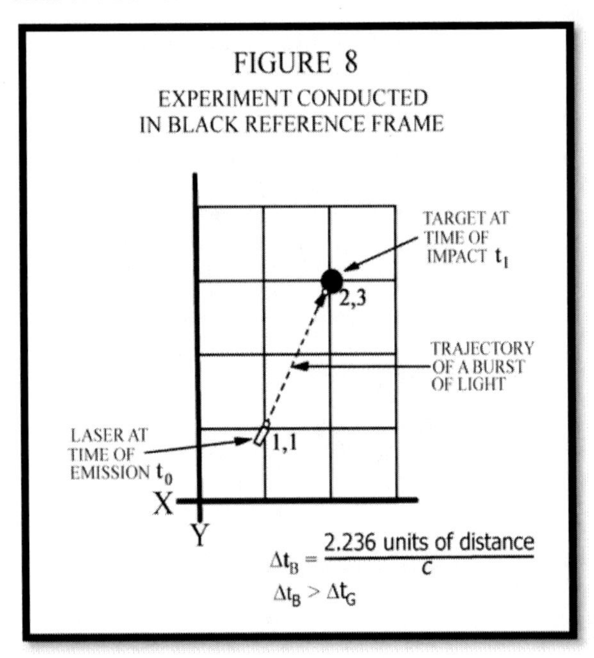

For illustrative purposes, it is assumed that both reference frames have identical coordinate systems and their X and Y axes are superimposed at the instant when the light is emitted.

As shown in Figure 9, the gray reference frame is moving upward and to the left relative to the black reference frame. A twelve-square section of each coordinate system is shown for reference. The X axis is at the bottom of each twelve-square section and the Y axis is at the left side. The location of the gray reference frame relative to the black reference frame is shown at three equally spaced instants in time, t_0, t_1 and t_2. For illustrative convenience, the interval of time between them is equal to the time required for the light to travel from the laser to the target in the black reference frame. At t_0, the origin of the gray reference frame is below and to the right of the origin of the black reference frame. At time t_1, their origins and axes are superimposed. That is the instant in time when the laser emits the burst of light. Each observer observes that event as occurring at coordinates 1, 1 in his reference frame.

At time t_2, the gray reference frame's coordinate system has moved upward and to the left to its top position in the illustration. At that instant, the burst of light strikes the target. The black reference frame's observer sees and marks that event in his coordinate system at coordinates 2, 3. The gray reference frame's observer sees and marks the location of that event at coordinates 3, 1 in his coordinate system.

The gray observer calculates that the light traveled two units of distance in his reference frame during the interval of time between the two events. The black observer calculates that the light traveled diagonally approximately 2.236 units

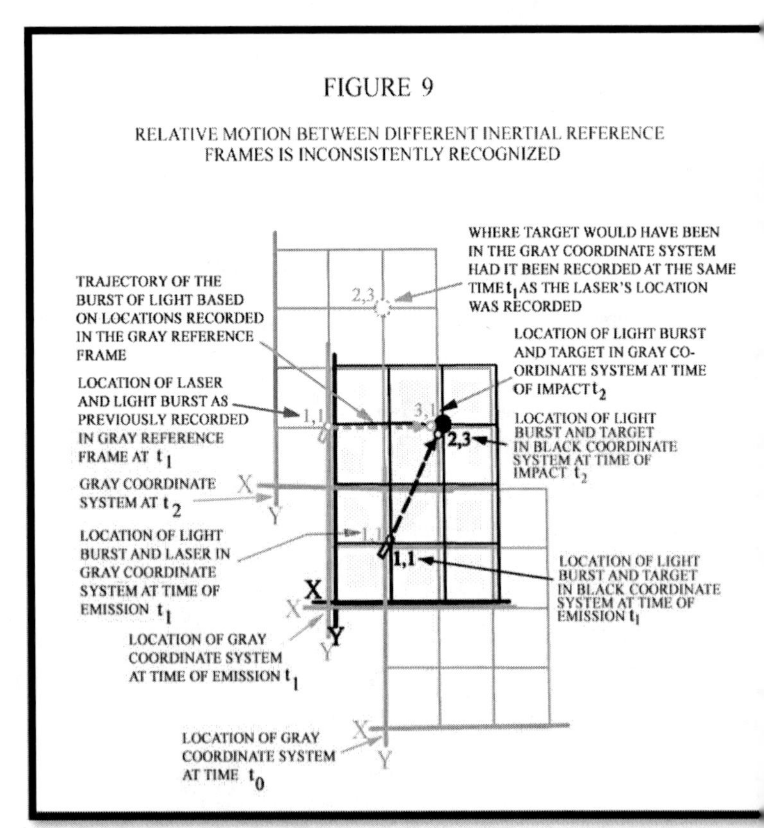

of distance in his reference frame between those same two events. Since the light traveled farther in the black reference frame between the same two events than it did in the gray reference frame, one must conclude that time passes more slowly in the gray reference frame than it does in the black reference frame.

One might stop at this point and declare a victory for the special theory. However, there is one little glitch to deal with. The only way the laser could remain at coordinates 1, 1 in the gray reference frame at time t_2 is if that part of the gray reference frame had remained stationary relative to the black reference frame during the interval of time between t_1 and t_2. But if it had done so, the entire coordinate system of the gray reference frame would have to remain stationary relative to the entire coordinate system of the black reference frame. And, if it had done that, the target would be at 2, 3 in the gray reference frame at time t_2 instead of being at 3, 1. Recording events which occur at different points in time in a different reference frame as if their locations were stationary in the observer's reference frame treats the relative motion between the coordinate systems of the two reference frames differently for some observed events than for others. This location recording error is innate to observations made in inertial reference frames of events which are separated in time and occur in another reference frame.

7. What an Observation is Believed to Mean Depends Entirely on its Interpretation

Figure 10 shows two spaceships which are stationary in two different inertial reference frames. The reference frames are named A and B. They are in motion relative to each other at a constant velocity v_d. v_d is the difference between their respective inertial velocities. Because each reference frame is in motion relative to the other, the observers will see the same celestial bodies but will perceive them as moving in different directions at different speeds. This is business as usual for the special theory. Observers in each inertial reference frame will see the celestial bodies in space moving in different directions at different speeds than observed by observers in other reference frames. According to the premise of equal merit, the motions of the celestial bodies actually are

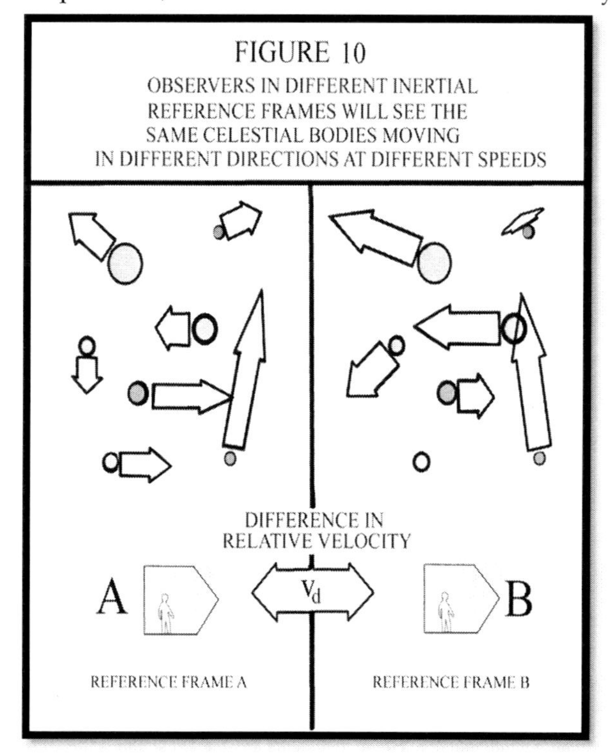

different in each different inertial reference frame.

Figure 11 takes a different approach to describing how the two different reference frames relate to each other. It shows a single spaceship in reference frame A on the left side of the illustration. When in reference frame A, the observer in that spaceship sees the same celestial display as shown for the observer in reference frame A in Figure 10. The observer in the spaceship then uses the spaceship's rocket to change its velocity by Δv to that of reference frame B. Note that Δv has the same magnitude as v_d. However, it now can be recognized as the change in velocity which is required to move from reference frame A to reference frame B. The distinction between a change in velocity and a difference in velocity is that the special theory treats a difference in inertial velocity as working the same way in both directions. A change is not subject to that ambiguity. If the spaceship observer changes the spaceship's inertial velocity horizontally to the right by Δv, as shown in the middle box in Figure 11, he will take his coordinate system and his human senses along with him. This will alter his perception of the motions of the celestial bodies he saw when he was in reference frame A. That alteration will be unique to him and totally subjective. The motions he observes in reference frame B will become the vector sums of what he observed when in reference frame A (the white vectors) and the change in his definition and perception of motion caused by his own change in inertial velocity (the gray vectors).

The net effect of their vector summation is shown in Figure 12. Not surprisingly, it is no different from what is shown in Figure 10, except for the explanation (Figures 11 and 12), or lack thereof (Figure 10), of what causes the difference in the observations. In Figure 10, the special theory defines both observers' observations as having equal merit.

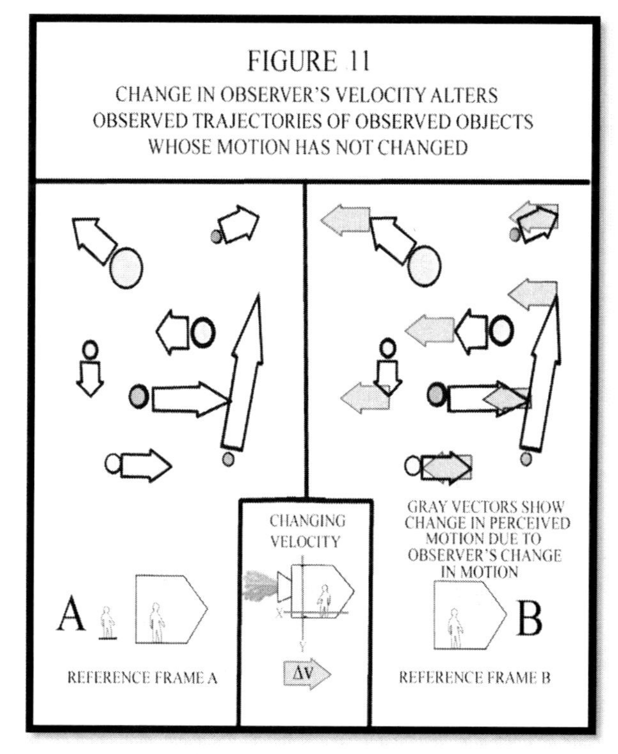

The motions of the celestial bodies actually are different in

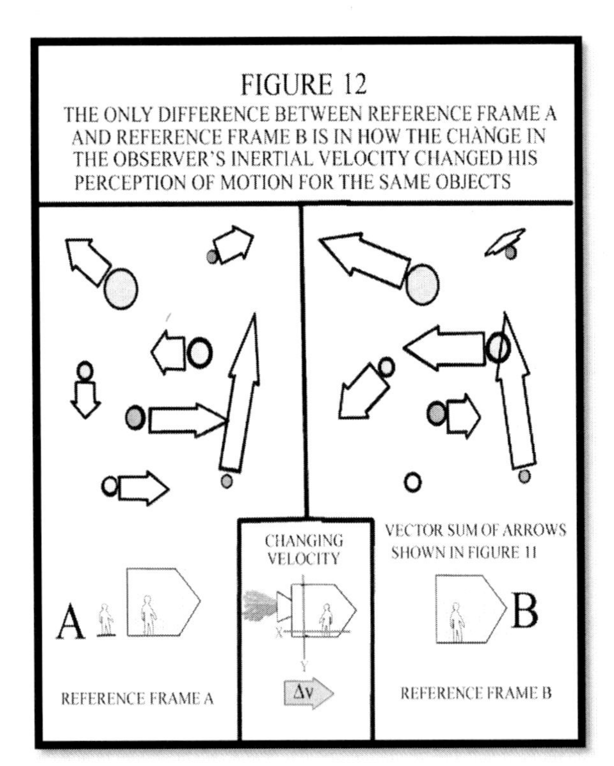

reference frame B than they are in reference frame A. However, as shown in Figures 11 and 12, the motions of the celestial bodies haven't changed; the observer has experienced a personal and entirely subjective change in his definition and perception of motion. The spaceship observer and his spaceship are the only things that have changed their inertial velocity. One would need an almost limitless amount of rocket fuel and a virtually uncountable number of rocket engines to change the motions of all of the visible celestial bodies in the universe.

Given the circumstances of this thought experiment, Occam's razor would suggest that the difference between the motions of the celestial bodies observed in reference frames A and B of Figure 12 is caused by simple observation error when the observer is in reference frame B. If the observations in Box A were correct, the observations in Box B must be incorrect. To conclude otherwise not only flies in the face of Occam's razor but requires a strong dose of magical thinking (yet another attribute of the human condition).

8. A Century of Validation by Circular Reasoning

Figure 13 shows how subconscious reliance on the special theory's own system of beliefs has influenced the design of experiments and the interpretation of empirical data which have consistently validated it.

The box at the left of Figure 13 provides an overview of the special theory's foundational beliefs. At the top are its postulates. Next are other premises which support, clarify or are derived from the postulates. These are the beliefs which must be valid in order for the special theory to be valid. At the bottom of the left-hand box are the theory's predictions about the nature and consequences of inertial motion.

The circles in the box on top show some of the key beliefs which guide how the theory reaches its predictions. The circles in the bottom box show the key steps involved in the design and conduct of experiments involving inertial motion.

The lines between the circles in the top box and those in the bottom box show where subconscious acceptance of the theory's own foundational beliefs can guide empirical analyses to a preordained validation. For example, the influence of the belief in circle 4 on experimental design is hard to dismiss. An experimental design based entirely on experiments conducted in and observations made from inertial reference frames clearly is based on the expectation that they will capture all of the data which are necessary to understand motion (circle 4) and that the resulting data will fully and accurately describe the phenomena being examined (circles 1 and 2). It also is clear that the beliefs in circles 2 and 4 play a decisive role in validating observations of the same events which provide conflicting information on the direction and distance traveled by the same emission of light (circle 6). Given the "validated" empirical data, the implicit beliefs in circles 2 and 3 leave no alternative to concluding that the theory's predicted time dilation, length shortening and increased mass are the only means for reconciling the empirical data.

Customary experimental practice appears to be governed by a subconscious belief that the concepts underlying the special theory are unquestionably valid. This subconscious conviction affects experimental design, validation of conflicting observations as having equal merit and interpretation of the resulting empirical data.

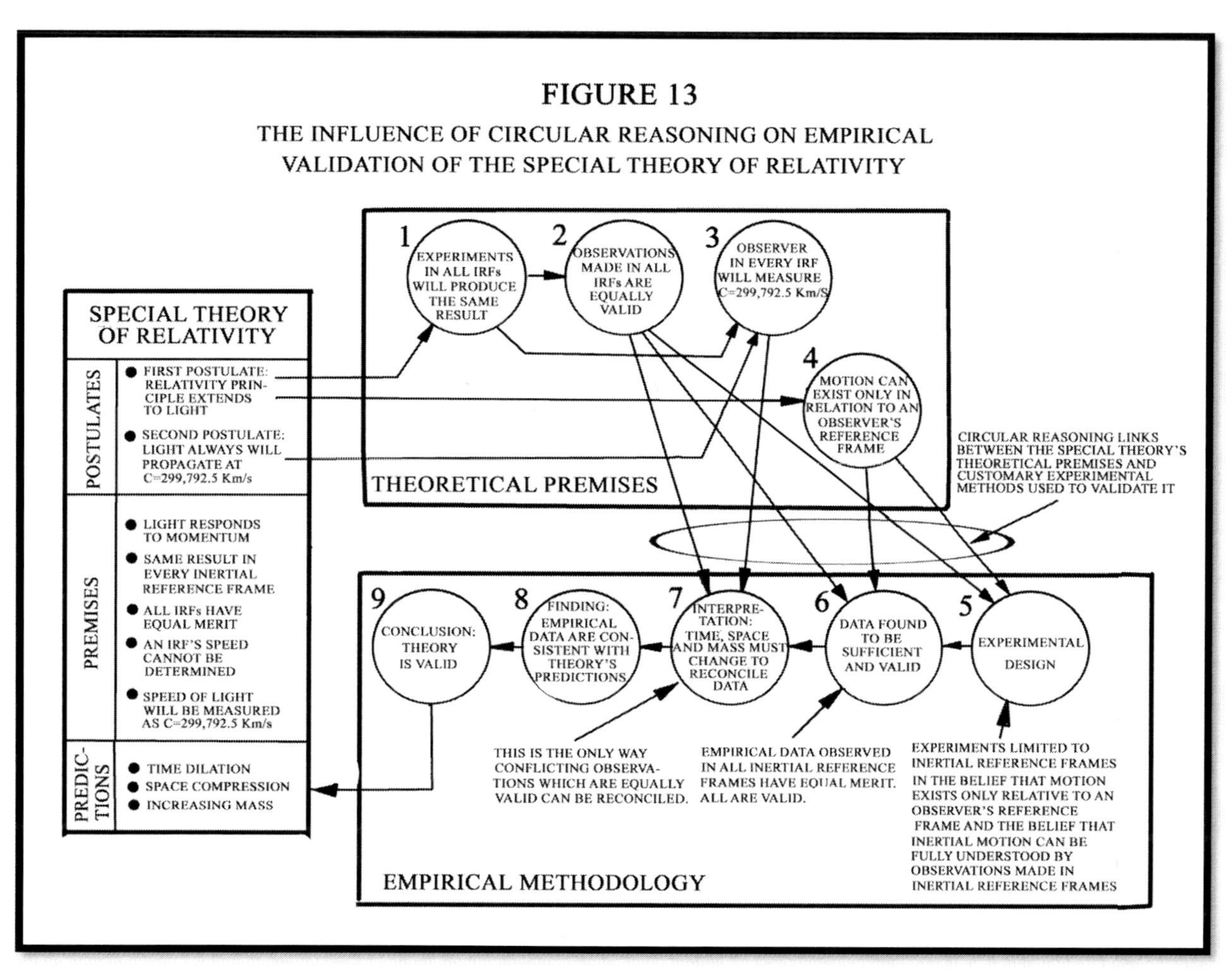

Experimental designs which address inertial motion are confined exclusivity to inertial reference frames (A comparison between how the theory treats a change in velocity with how it treats a difference in inertial velocity between the same two reference frames can be made only by considering what happens in a non-inertial reference frame).

Conflicting observations of the direction and distance traveled by phenomena under study are accepted as having equal merit (the observations were made in inertial reference frames). Interpretation of the empirical data begins with an assumption that the phenomena being observed behaved in accordance with the postulates and premises (e.g., it is assumed that the light responded to momentum in accordance with the difference in its source's velocity relative to the observer).

Given the above influence on experimental methodology, it is not unreasonable to expect that the explanation of any differences in the empirical observations can be found only in the theory's predicted environmental conditions (time dilation, space compression and increased mass). The conclusions to be reached by such an empirical methodology are virtually inevitable. The theory consistently will be affirmed.

However, as shown in Figures 3 and 6, if the postulates are correct it is the change in velocity required to move from one reference frame to another which changes the observed phenomena's behavior. Because of the circularity between an observer's state of motion and his definition of motion, observers in different inertial reference frames will disagree with each other because of observation error. Each observer has his own entirely subjective definition of motion which is different from the entirely subjective definition of the other. Whichever of the two is correct, the other must be incorrect.

9. Conclusions

What this paper shows is that two changes in the experimental design of Dr. Einstein's famous thought experiments will cause the special theory's postulates and premises to contradict each other. The two changes are:

- The omnidirectional light source available in the early 20th century is replaced with a directional light source, the optical laser.
- A distinction is made between a physical reference frame, such as a spaceship, and an inertial reference frame, which is simply a state of motion.

Using a laser as the light source imposes the discipline of having to recognize that the propagation of light is a definite, constant vector phenomenon, not simply a definite, constant scalar speed. That eliminates the ambiguity over the trajectory of the same unit of light which is being observed simultaneously from different reference frames.

The distinction between a physical reference frame (e.g., a spaceship) and its state of motion (i.e., an inertial reference frame) permits the same experiment to be made in one inertial reference frame and then be moved, intact, to another. Both the magnitude and the vector direction of the spaceship's change in motion are defined. While the states of inertial motion of the two reference frames may not be known, the change in motion from one to the other is absolute, not merely relative. The two postulates of relativity then can be used to determine what change must occur in the laser's trajectory in the second reference frame relative to what it was in the first one. That, in turn, permits a comparison between what the postulates say must happen and what the spaceship observer will observe. What that discloses is that if the observation made in the first reference frame is correct and the postulates of relativity are correct, the spaceship observer's observation in the second reference frame will be incorrect. He will fail to detect the change in the light's trajectory due to the change in the spaceship's momentum. That is because the coordinate system he uses to define locations has changed its own state of motion to match that of the spaceship, observer, laser and target. The method of defining motion by means of changes in location has been corrupted by making the system used to identify locations subject to the observer's state of motion. Such a circular method of measurement is innately meaningless.

Based on the results presented in this paper, the following conclusions can be made:

1. The special theory of relativity is a house of cards based on a parlor trick and sustained by circular reasoning.
2. When a laser's own inertial motion is changed, the change in its trajectory which is required by the postulates of relativity cannot be detected by an observer who is in the same physical reference frame as the laser. That observer will suffer from observation error caused by the change in his own state of motion and the corresponding change in the coordinate system of his and the laser's reference frame. Thus, the premise of equal merit is refuted by the very postulates upon which it has been based.
3. Because the premise that light always will be measured to travel at its definite speed c depends on the premise of equal merit, the premise of will measure c also is invalid. The speed of light is constant, but what is observed depends on the observer's entirely subjective definition of motion.
4. Because both the premise of equal merit and the premise of will measure c are invalid, the special theory and its predictions are invalid.
5. Physical objects provide the same observed results in all inertial reference frames because a change in their momentum changes both their trajectory and their speed in the same proportion. However, the second postulate requires light to propagate through empty space at the same constant speed c. Thus, even if light responded to momentum, its change in trajectory would not be matched by a corresponding change in its speed. The interval of time between when a unit of light is emitted and when it arrives at the target would change as the reference frame's state of inertial motion changes. *Experiments involving the propagation of light can provide the same apparent trajectory in different inertial reference frames but cannot produce the same time interval.* Thus, the second postulate of reletivity refutes the first postulate. The results of experiments made in different inertial reference frames will be different. The special theory is rendered invalid by conflicting requirements at its very foundation.
6. Because the special theory and its predictions are invalid, time, space and mass are restored to their historical status as the universal constants of physics (i.e., time does not slow, length does not shorten and mass does not increase with motion).
7. Because neither time nor space are affected by motion, the relationships described by the special theory which make them appear to be interchangeable are invalid. The concept of space-time is inoperable.
8. Mass and energy are different in kind and are not interchangeable. Mass is a scalar phenomenon and is inert until pushed by energy. Energy is a vector phenomenon which changes the state of motion of mass.Their innate characteristics and their effects on their environment are polar opposites.
9. Because the first postulate is invalid, there is no basis for maintaining that light, which has no mass, responds to momentum.
10. Light clocks are inaccurate at high inertial velocities. Mechanical clocks are not.
11. Because string theory depends absolutely upon the validity of the special theory of relativity as written, string theory is DOA.[xxxii]
12. The belief that the motions of celestial bodies actually are different when being observed from different inertial reference frames is invalid. That belief is based on the invalid premise of equal merit. There are no multiple universes.
13. Interpretations of experiments involving the motions of celestial bodies over the past century need to be reexamined.
14. The human condition has significantly undermined the scientific method for the past century. Its influence on scientific decisionmaking needs to be better understood and guarded against.
15. Empirical observations do not constitute proof. All they can do is indicate what might be true subject to the adequacy of experimental design, the capabilities of available technology and the validity of interpretations of whatever empirical data happened to be detected. This is not a critique of the scientific method; it is an unavoidable consequence of the human condition.

---

[i] Douglas G Giancoli, *Physics*, 4th edition (Englewood Cliffs, New Jersey: Prentice Hall, 1995), 744-745.
[ii] Ibid., 743.
[iii] Ibid., 750. Goldsmith, Dr. Donald, and Robert Libbon, *Einstein: A Relative History* (New York: Simon & Schuster, Inc., 2005), 70.
[iv] Cox, Brian and Jeffery Forshaw, *Why Does* $E=MC^2$*: (and why should we care?),* (Cambridge, Massachusetts: Da Capo Press, A Member of the Perseus Books Group, 2009), 39-45. Giancoli...,. *Ph;/'ysics*, 745. Gribbin, John and Mary Gribben, *Annus Mirabilis: 1905, Albert Einstein, and the Theory of Relativity* (New York: Chamberlain Bros., Penguin Group, Inc., 2005), 96-97.
[v] Giancoli, *Physics*, 744-745. Goldsmith, *Einstein: A Relative History*, 48. Gribbin, John, *Annus Mirabilis:* 96-97.
[vi] Goldsmith, *Einstein: A Relative History*, 67-70.
[vii] Knight, Randall K., *Physics for Scientists and Engineers: a strategic approach (San Francisco, California*: Pearson Publishing, Inc., publishing as Addison Wesley, 2004), 1149.
[viii] Giancoli, *Physics*, 744-745.
[ix] Ibid., 743.
[x] Perkowitz, Sidney, *Empire of Light: A History of Discovery in Science and Art* (New York: A John Macrae Book, Henry Holt and Company, 1996), 61.
[xi] Cox, Brian, *Why Does* $E=MC^2$, 28, 41. Goldsmith, Dr. Donald, *Einstein: A Relative History*, 49. (Note that the value for the speed of light *c* is expressed as 299,792,458 m/s, 299,792.5 km/s, or 3.00x10^8 m/s depending on the reference source. All are essentially the same number with only slight differences in rounding.)
[xii] Cox, Brian, *Why Does* $E=MC^2$, 28.
[xiii] Giancoli, *Physics,* 745. Perkowitz, *Empire of Light*, 61-63.
[xiv] Giancoli, *Physics,* 745-746.
[xv] Aczel, Amir D., *God's Equation* (New York: Four Walls Eight Windows, 1999), 22. Cox, *Why Does* $E=MC^2$, 29. Giancoli, *Physics,* 745.
[xvi] Giancoli, *Physics, 745.*
[xvii] Perkowitz, *Empire of Light*, 65-68. Giancoli, *Physics,* 746-749.
[xviii] Giancoli, *Physics*, 749.
[xix] Aczel, *God's Equation*, 24. Rigden, John S, *Einstein 1905: The Standard of Greatness* (Cambridge, Massachusetts: Harvard University Press, 2005), 84. Giancoli, *Physics*, 751. Hey, Tony and Patrick Walters, *Einstein's Mirror,* (New York: Cambridge Press, 1997), 7.
[xx] Giancoli, *Physics,* 750.
[xxi] Ibid. Goldsmith, *Einstein: A Relative History*, 67.
[xxii] Giancoli, *Physics,* 750. Goldsmith, *Einstein: A Relative History,* 70.
[xxiii] Ibid.
[xxiv] Giancoli, *Physics,* 750.
[xxv] Giancoli, *Physics,* 743-744.
[xxvi] Hey, *Einstein's Mirror*, 43. Goldsmith, *Einstein: A Relative History,* 67.
[xxvii] Giancoli, *Physics,* 750.
[xxviii] Goldsmith, *Einstein: A Relative History*, 67.
[xxix] Ibid., 48. Gribbin, *Annus Mirabilis,* 96-97. Giancoli, *Physics.* 745.
[xxx] Goldsmith, *Einstein: A Relative History*, 49.
[xxxi] Giancoli, *Physics,* 166-167.
[xxxii] Smolin, Lee, *The Trouble with Physics; the rise in string theory, the fall of a science, and what comes next* (New York: Houghton Mifflin Company, 2006), 223.

Light, Gravity, and Mass

A Particle Theory

Robert & David de Hilster
Boca Raton, Florida, United States
e-mail: robert@dehilster.com, david@dehilster.com

The electromagnetic spectrum including visible light is most often characterized by wave theory. But there are effects that indicate it has particle characteristics. Hence, there is an ongoing discussion of wave particle duality. There has been work trying to make a wave act like a particle. This paper uses a particle and provides a mechanism for it to have wave properties. If it is a particle, could the same particle that is the cause of EM radiation also be the particle needed for pushing gravity? And how does mass fit in?

1. Introduction

(David de Hilster) For decades, scientists have been trying to find a model of light and gravity both separately and togaether. For this paper, we are assuming the following:

1. There is only space and mass and all mass is in movement
2. Electrical fields, gravity, light, and everything else in the universe are made up of mass in motion and are pushes
3. Light and gravity both travel at the velocity of c.
4. Electric fields are matter in motion

Evidence for light and gravity traveling at velocity c come from the Wang experiment [1]. The Wang experiment shows clearly that the effect of gravity and light from the sun arrive at the same time.

In order to adhere to both light and gravity traveling at the speed "c", the particle theory of waves is proposed.

2. The Particle Theory of a Wave

(Robert de Hilster) One particle cannot describe a wave. But a group of particles could be arranged such that a simple or complex wave is described. If the particles are distributed such that they are close togaether in the first part of the wave and farther apart in the second part of the wave, the distribution could resemble a sine wave. This is illustrated in Figure 1.

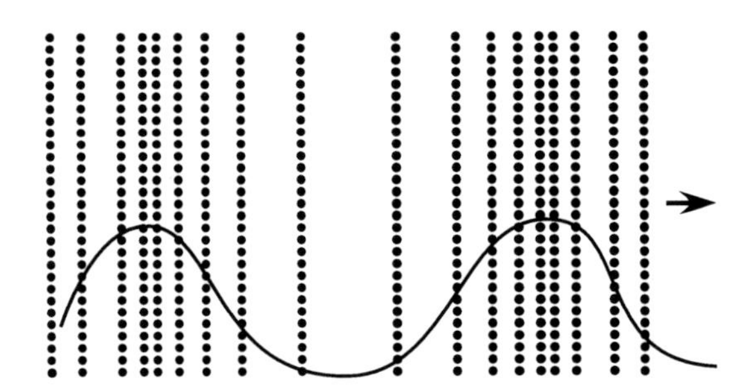

Figure 1 -Encoding Waves with Particle Density

The above dot graphic is an idealized visualization of the particle theory of a wave.

3. Sine Wave

A sine wave is shown as a continuous mathematical model. The mathematics of a sine wave is shown in equation 1.

$$a_t = A \sin 2\pi f t \qquad (1)$$

There are only two terms in this equation. There is Amplitude and frequency. Equation 2 describes the frequency of this wave in terms of wave length λ and velocity, v.

$$f = v/\lambda \qquad (2)$$

Equation 1 and 2 are mathematical equations for which there is no physics. It was Max Planck [2] that defined the energy of a charged atomic oscillator as being proportional to the frequency of a wave. This is Planck's equation:

$$E = hf \qquad (3)$$

Although equation 3 appears continuous, the energy levels are reported as quantized.

4. Particle Wave

The physics for the particle theory of a wave is described in section 2 and can be shown as a sequence of digital values.

(55, 60, 70, 60, 55, 45, 40, 30, 40, 45) **(4)**
(500 particles in a period of one second)

There are 300 particles in the first half and 200 particles in the second half. Each number represents the number of particles in 0.1 seconds. The specific shape of this wave is shown in a series of flat steps which indicates distortion, but the basic wave and frequency are still there.

Amplitude

The amplitude of the wave shown in sequence 1 is determined as follows:

$$A = (D_{Max} - D_{min})/2 \qquad (5)$$

The amplitude is A. The Density at peak is D_M, and the Density at Minimum is D_m. The value of A for sequence 4 is 20 particles. If the distribution is a constant density, then there is no amplitude or frequency.

Frequency, Wavelength, and Velocity

The frequency of the particle wave is determined by the amount of time that the peak value takes to repeat. In sequence 4, this is one second and so the frequency is one cycle per second. The velocity of the particle is designated v_p. The equation for the frequency of the particle wave is:

$$f_w = v_p/\lambda_w \qquad (6)$$

Intensity

For the particle wave, the intensity is different than amplitude. The amplitude helps define the frequency. The intensity of the wave is related to the energy. In sequence 4, intensity is proportional to the number of particles per wave.

$$I \propto N_p/\lambda_w \quad (7)$$

Using equation 6 to solve for λ_w we get:

$$I \propto \frac{N_p}{v_p} f_w \quad (8)$$

Intensity is proportional to frequency. But the number of particles per wave is an integer number and so equation 8 clearly shows that the value of intensity is quantized. If this model is true, then Planck's constant is not a constant!

5. Applying the Theory

Light and Gravity

It seems logical that this theory can be applied to the full electromagnetic spectrum, from Gamma rays to Radio waves. The highest frequency of gamma rays can be generated if enough particles can be put in a very short wavelength. But physical restrictions will limit the upper frequency.

But it would appear that there is no restriction at low frequencies. In fact, a stream of particles could have zero amplitude and zero frequency. A stream of particles like this is exactly what is needed for pushing gravity. Could this one particle be valid for light and gravity? That's part of what we are trying to find out.

Mass

There is nothing in this particle theory that tells us about mass. It was Newton who defined the value of mass as the ratio of its acceleration relative to the acceleration of a standard mass. That's a definition of its value, not a definition of what it is.

Mass is known when its motion is detected. We know the value of the mass of the moon because we see it move. We cannot see the gravity particle move therefore the value of its mass is unknown. We don't see the light particle move either, and its mass is also unknown.

6. Opportunities

When you have a new theory, there are many questions that need to be addressed. Here are some ideas that need to be explored.

Speed of the Particle

The speed of light is generally understood, but not completely. What causes light to slow down in water and glass? If light is a particle, then drag seems to be a possibility. The real hard part is what causes the light to speed up when it moves from water to air?

Do gamma rays and radio rays speed up and slow down? If gravity is caused by the same particle does it speed up and slow down?

Interaction

The interaction of this particle with other objects is quite varied. Here are some of the interactions and all need to be explained. The particle can:

Reflect (one or many frequencies)
Refract (change direction)
Pass through objects
Attenuate (lose some particles)
Absorb (be captured)
Push (cause objects to move)

Depending on the frequency, amplitude, intensity, and speed of the wave any of these can happen. The theory needs to be applied to gamma rays, visible light, radio waves, and gravity.

There is a lot to do!

7. Particle Interaction

Here is one model for the particle interaction with an object. Figure 2 shows this particle entering a water molecule. It has two possibilities, stay with the molecule or be ejected by the water molecule.

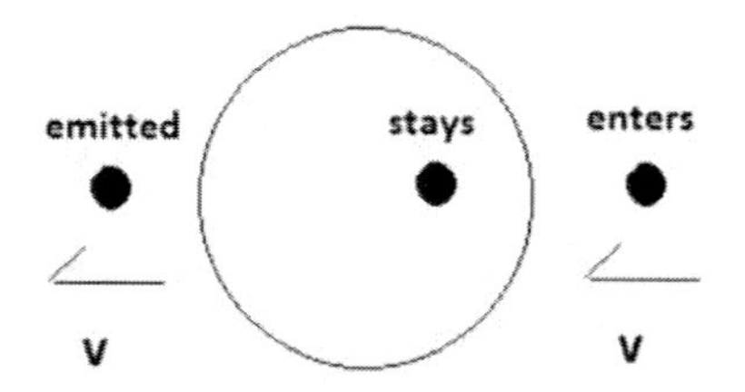

Figure 2 -Particle Entering/Leaving Water

Stays

In this case, the particle enters the water and stays. The manner in which it stays is not known. The particle pushes the water and it is added to the amount of matter. This satisfies the requirement that gravity is a pushing force and suggests that this process may be part of the earth's expansion.

The mass of the particle in a wave is unknown. But if it is captured by the object, it adds matter. This added matter can now add to the mass of the object. Regardless, the mass of this particle is still not known.

Emitted

The second option is that the particle is re-emitted, or a different particle is emitted, and is emitted at the speed of c (the speed of light in a vacuum). If the particle pushes the object forward as it enters, then the object is pushed backward when it is emitted. In this case the pushing of the object is cancelled.

The release of the particle is assumed to be controlled by forces at the atomic level. Due to cause and effect, the emitted particle leaves the water molecule a fraction of a second after it entered. So between molecules, the speed of the particle is c. But the short time delay in the molecule causes the overall velocity of the particle to slow down. It is well known, that the speed of light in water is slower than the speed of light in air.

When the water particle is emitted from the last water molecule, it is emitted into the air at speed c. It was emitted at speed c when it started light years ago, and it is emitted here the same way. The particle enters the air at c, but soon slows down due the delay through the air molecules. In any case, the speed of the particle was slow in the water is now faster in air. This is how the light wave **speeds up** going from water to air.

8. Metals, Magnetic Fields, and Electricity

(David de Hilster) Given the assumption that there is only mass and mass in motion in the universe, magnetic fields and electricity have to also be mass in motion.

Ionel Dinu's underwater cylinder experiments [3] show a clear model for electric fields. This clearly shows that electric and magnetic fields are spinning mass.

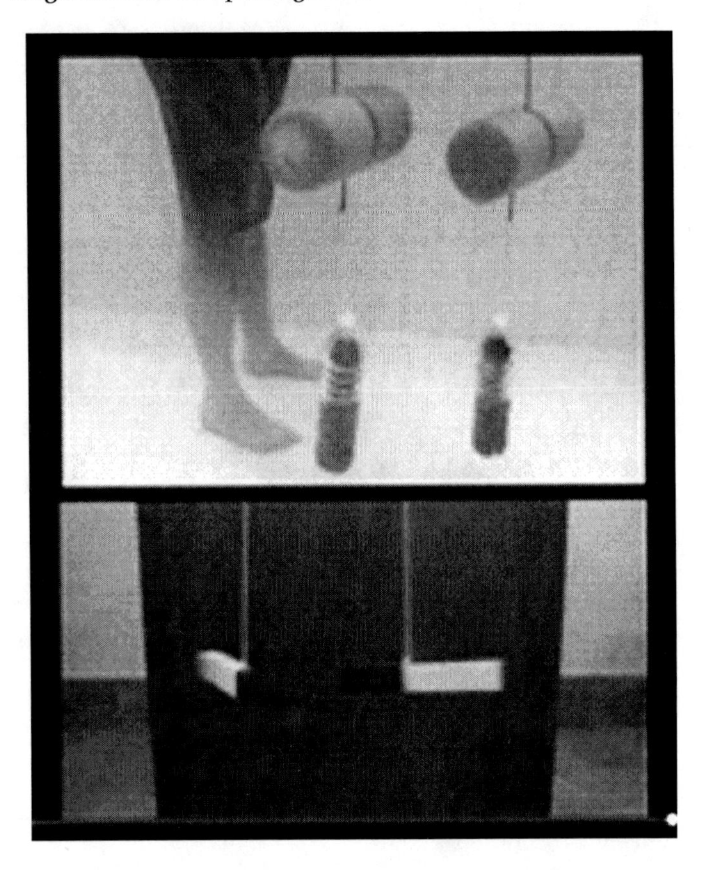

Figure 3 - Spinning masses underwater act as magnets

With the assumption that magnetic fields are swirling mass, we can conjecture that the whirling mass is gravity or aether particles.

Curiously, this happens only in metals. Why would the structure of metals affect aether which other materials do not? If electricity is light as proposed by Borchardt and Puetz [4], this again points to special properties of metals and their ability to interact with aether.

9. Antennas

This model of gravity and light also speak to interacting with aether to create varying densities in aether flow. Again, this happens in metals. How does this happen?

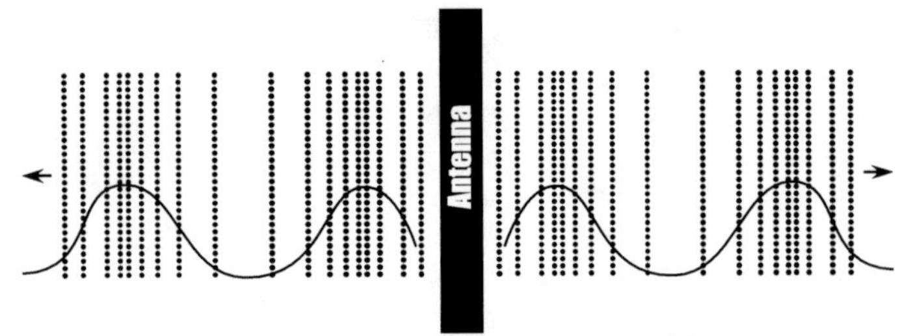

Figure 4 - Waves or density changes created by antennas

10. Interaction with Particles and Mass

This model of gravity and light being made of aether must be extended and adhere to the behavior of we observe. For instance, in a cave, gravity diminishes very little but certain electromagnetic waves are filtered out. How does this happen? With this model, this means that the number of aether particles remains the same but the density changes are eliminated for certain wave lengths.

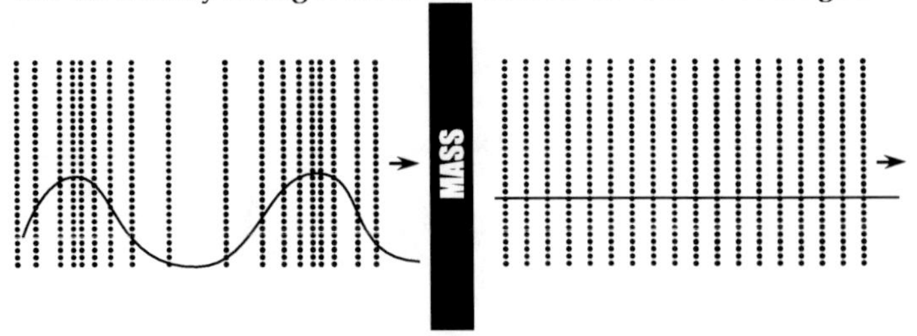

Figure 5 - Waves being filtered by mass

11. Lasers Are Explained

This model explains how laser light could exist. It is simply a concentrated group of particles traveling togaether. Whereas normal light sources radiate out from a source and the particles diverge, lasers emit particles that travel without and travel in the same direction.

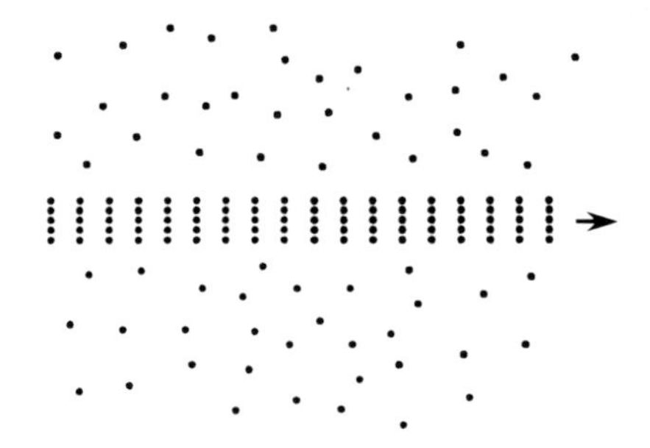

Figure 6 - Laser light as particles traveling togaether

12. Summary

(Robert de Hilster) This is an interesting model but needs experiments to prove it is a valid model. Only when it meets all or most requirements can the model be considered useful. Today there is only indirect evidence of gravity. Motion is indirect evidence. Only when we have direct evidence will we be able to confirm the model. We need to be able to see the particle doing all these things,

References

[1] "Want Eclipse", Robert de Hilster, 2010, NPA Proceedings.
[2] "Planck's Constant", https://en.wikipedia.org/wiki/Planck_constant
[3] "Ionel DINU - Unravelling the NATURE of the MAGNETIC FIELD", Youtube video: https://youtu.be/RNnXVKyk-kA.
[4] "Universal Cycle Theory", Steve Puetz and Glen Borchartd.

Gravity is A Pushing Force

This Gravity Subgroup – Wave Theory (GSG-WT) is a member of CNPS's Gravity Interest Group (GIG)
Members of the GSG-WT are:
Paul Schroeder, Arthur Ramthun, and Robert de Hilster.
4/20/2015

This document proposes that gravity is caused by an Electromagnetic EM wave that pushes an object as it passes through. There are other models that are based on pushing theories, but the presentation here suggests that the Electromagnetic Wave theory is the most appropriate.

Table of Contents

1. History

There have been many attempts to produce theories that define and explain the nature of gravity. Newton had no theory but developed a very good equation for gravity. Einstein had a theory based on space time bending which also doesn't provide a nature for gravity. The actual assignment of a physical nature to gravity begins with Fatio and Le Sage theories based on a pushing particle. Later on Maxwell introduced the idea of a gravity wave using mathematical equations. Our paper provides a theory in which gravity exhibits the properties of electromagnetic waves.

2. Pushing Force

Fatio [1] and Le Sage [2] introduced the idea of gravity as a pushing force potentially to overcome the 'action at a distance' implications of a pulling force as assumed for Newton's gravity [3]. Logically 'the horse does not pull the wagon it pushes the harness. There are many examples of a pulling force that can actually be explained using a pushing force.

Figure 1 is a drawing made by Le Sage that shows his particles entering a spatial object and then leaving the object diminished and containing fewer particles. Therefore the pushing particle density in between two objects is less. There is less pressure between the two objects so the pressure on each is less than the pressure from outside of the shadow. Hence there is a 'net' force pushing the objects together. For these objects and specifically for us on earth, the pushing particles must come at us from all directions in order to have gravity push us toward the center of the earth no matter where we stand.

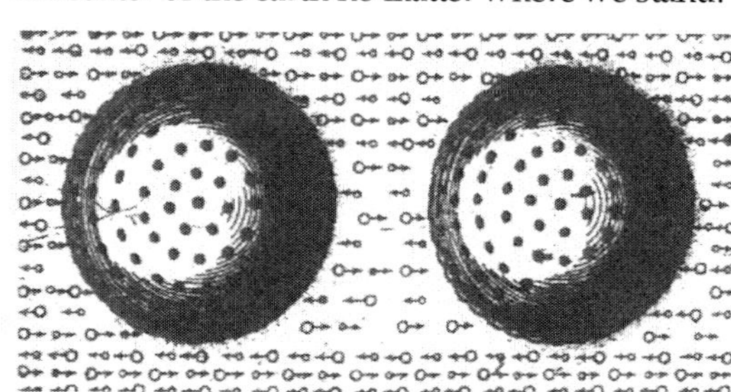

Fig. 2. Pairs of macroscopic bodies traversed by currents of ultramundane corpuscles. From Le Sage's *Essai de chymie mécanique*. Photo courtesy of the Library of the Royal Society, London.

Figure 1 –Particles Pushing the Objects together

In the model we advance here the particles in figure 1 are replaced by Electromagnetic waves. To proceed, these waves must likewise be able to push the object. So how does an electromagnetic (EM) wave push? This paper intends to answer this question.

3. Assumptions

Assumptions are used to help establish the context for explaining the theory. These assumptions may never be proved, but they must not contradict any part of the theory.

a. **Space:**
   1) Space is where objects exist.
   2) Space extends in 3 dimensions and is similar throughout.
   3) Space is not equivalent to, nor does it contain, any void. Any void area would be immeasurable and therefore not exist.
   4) Space therefore has structure, one possible example is plasma.

b. **The Structure of Space**
   5) The structure of space is in constant motion and simulates a material nature.
   6) Space structure is in equilibrium unless the motion of its contents is interfered with.

c. **Motion**
   7) Time is man-made idea for measuring motion.
   8) Motion is a function of both matter and waves.
   9) A net' directional amount of motion can occur within the structure of space. That motion becomes detectible, and can apply pressure on other matter.

10) Motion is detected when an object's position in space changes from one instant of time to another.
11) Significant matter, often called a mass, interferes with the motions of the structure of space.

4. Definitions

d. Gravity

12) Gravity Source - Our pushing gravity is the source or 'cause' of the pressure that produces detectible effects. This source is currently undefined and has been since the time of Newton.
13) Gravity Effects - The term gravity is often used to denote the detectable 'effects', such as the falling apple. The effects of gravity upon matter bodies results in detectible motions or pressure. The 'effect' is the 'attraction' of matter toward mass bodies. The attraction effect is expanded by Newtonian dynamics to include the potential motion of two or more spatial bodies toward each other.

e. Mass/Matter

14) Matter is normally used as a term to distinguish visible or sensible things from apparently empty regions. The existence of regions which have only waves is often incorrectly referred to as void.
15) Mass is a property of matter which modifies gravitational pressure (causing motion).

f. Particle.

16) A gravity particle is an object in motion that applies pressure upon contact.
17) Gravity is caused by 'particles applying external pressure' (PAEPs). This term is coined to avoid calling the gravity particle by the term 'graviton'. The actual particles may be matter or radiation waves.
18) Not only electrons but all other particles, charged or uncharged, show wave-like characteristics. [4]

g. Wave

A wave is a single, but repeating, non-linear flow within a ray or a beam.

h. Electromagnetic Wave

19) Light is an example of a transverse wave. A transverse wave is a moving wave that consists of oscillations occurring perpendicular (or right angled) to the direction of energy transfer i.e. the direction of propagation of the wave. A wave on a string is easily visualized as transverse waves.
20) For motions through space, a transverse wave is an oscillation relative to the direction of the beam and occurs in three dimensions. Oscillations in three dimensions emulate a coil.
21) Waves can be longitudinal. This means that there is some variation within the beam as it travels forward. The variation is repeated compressions followed by decompression along the length of the beam. The gravity ideas we develop using transverse waves can be adapted to compression waves.
22) To picture a wave in three dimensions, consider a particle revolving within the straw. From any direction the wave will look like a sine wave if viewed in two dimensions.

5. Properties of EM Radiation

An electromagnetic wave has certain properties described below.

i. Source

The main source of EM radiation is from the stars. This means that all the stars in the universe are a source of gravity. This implies that gravity comes at us from all directions.

j. Speed

The speed of EM radiation when serving as gravity duplicates the speed of light as denoted by c. c is the constant of speed dependent upon the medium of travel. If gravity is caused by EM radiation, then the speed of gravity is about the same as the speed of light. There is evidence of this from measurements made during a solar eclipse.

k. Frequency

There are frequencies of EM radiation that can easily pass through objects and there are others that do not. It is the long wave length, low frequency EM waves that can penetrate matter the furthest, even through spatial bodies. Energy is proportional to the frequency. So, some frequency is lost during interactions.

l. Amplitude

The amplitude of the wave is the radial displacement from the wave center. When waves push on objects some of its energy is 'used'. Energy is proportional to the square of the amplitude. and some of its amplitude is lost during an interaction. Figure 3 and [5]

m. Intensity

Radiant intensity is a measure of the intensity of electromagnetic radiation. We expect it is dependent on the number of beams per unit area. It is defined as power per unit solid angle. The SI unit of radiant intensity is watts per steradian ($W \cdot sr^{-1}$).

n. Applies Pressure

Upon contact with matter, EM radiation is found to apply pressure to that matter. There was uncertainty about this happening until Einstein assumed that light travels through space in concentrated bundles called photons. The photoelectric effect [6] revealed that light falling on a metal plate can liberate photoelectrons. The resulting equation is: $hv = E0 + Kmax$. [7] So an electron is pushed out and carries kinetic energy (Kmax), the amount of which is a function of the frequency. There are other cases of known pressure from light such as pressure on cones within the eye [8] providing vision and pressures upon earth's atmosphere by sunlight.

o. Phase

Waves come from all directions and from various distances. Any and all phases of EM waves may arrive at any point. The phase of the individual EM wave is not a consideration.

6. Actions of EM Waves

p. Source

Solar radiation is mostly light, with its associated wave length and frequency, for which we assume the amplitude, has been reduced. From all other directions the radiation is mostly long wave, so much so that we may not easily identify the beams as EM radiation. However this radiation retains full amplitude as it travels. At any point, as the beams continue flowing, the energy is retained as other beams enter the region from all directions. Essentially the ongoing merger of beams retains the normal pressure at all remote points in space. It is when nearby lower energy beams from bodies such as the sun cause an imbalance of motions within the structure of space that we experience 'net gravity'. The centripetal force attracting earth toward the sun is an example of this 'net gravity'.

q. Fabric of Space

The farther radiation travels from its source the greater the distance between rays. That gap is filled by rays from other sources. Over many years, Scientists have made the assumption that if there is a wave moving through space there must be a medium that allows that wave to propagate. But with all these rays from every direction defining space there may not be a need for additional mediums such as aethers. Space is filled by EM radiation which possibly becomes both the action and the medium. We call this the fabric of space. Essentially the light waves are examples of the movement within the medium just as sound is the movement of air.

Science has observed that light moving in one direction does not affect light moving in other directions when they intersect. The same non-interference could then apply to all EM radiation. While the radiation moves the interaction of waves from all directions gives the impression of stasis, or no detectible motions. Space can appear void while moving internally in all directions, even at speed c.

r. Moving Through Space

We will visualize EM waves in space as transverse. Being three dimensional we must picture a wave as a coil. A sequence of coils emulates the spring in a pen. The whole spring is moving rapidly at speed c suggesting its forward flow of the waves/coils be identified as a beam. Beams/rays individualize radiation fields just as particles do for matter.

s. Penetration

The waves are coils whose frequency determines whether a wave impacts matter mostly at the surface or within the mass. A beam with low frequency waves arrives more like an arrow and penetrates the mass. High frequency impacting waves may apply greater surface pressure. However, the total of surface pressure is minor relative to the penetration pressure of long waves which continues contact throughout the mass. Penetration contact applies throughout the internal body of mass. Matter is considered as mostly empty except for the nucleus. Figure 2 depicts the concept of penetration.

TRANSVERSE WAVE PENETRATION / INTERACTION MODEL

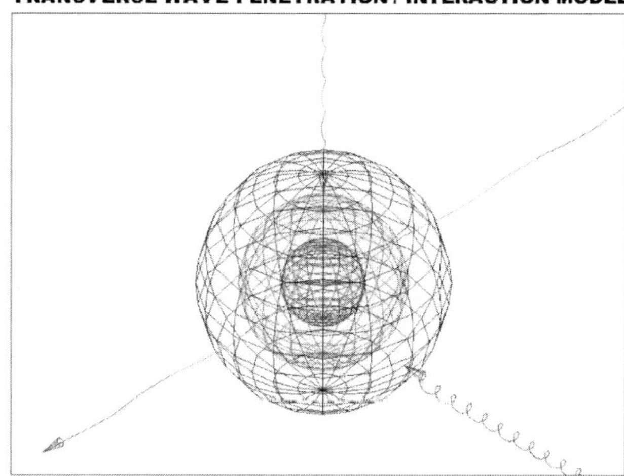

Figure 2 - Penetration

Figure 2 shows three different EM waves interacting with a spherical object. It is generally understood that waves of different high frequencies may either reflect off of the object, or penetrate a short distance into the object. Longer wave rays may pass through the object. Each incoming wave is shown in three dimensions and is depicted as a coil. The wave on the lower right does not penetrate, it reflects (not shown) and loses energy. The wave from the top enters a short distance and is absorbed by the object. Finally, the wave from the upper right passes all the way through the object. Each beam incurs interactions causing the amplitude and frequency to decrease.

t. Interaction

To understand EM gravity, consider motion that occurs when two billiard balls hit. The collision is a clear example of a mechanical interaction. The motion of one ball is transferred to the other in keeping with Newton's laws. This is an inelastic interaction in which all motion is retained by the balls externally. An elastic interaction allows some of the impact to be transferred to the interior of the impacted ball so the observable motion is only part of the result. With elastic interactions, some of the kinetic energy is converted to heat and thus lost. The creation of gravity 'effects' cannot occur with strictly inelastic interactions. To anticipate elastic interactions we focus on waves rather than mass particles being gravity sources.

The movement of a paper clip using a magnet indicates an interaction of the clip with the magnetic field which is an EM interaction.

u. Pushing

Throughout history mankind has studied motion, analyzing both the object in motion and the source of the motion. The object was matter. Being heavy it was assigned weight, was given the attribute of energy and became mass. The simplest cause of motion occurs from a push as a transfer of energy from one moving object to another object. Two of the events that cause motion - magnetism and gravity - did not follow that cause rule so the concept of force was invented. Spatial entities were originally thought to be massless objects. From Kepler forward stellar objects had mass and so Newton proposed the void where friction didn't interfere with the motion of objects. But the source remained the mythical - force.

The concept of light transmitting at speed 'c' violated rules as light did not seem to have any weight, and yet it had motion. So light moved as waves but lacked kinetic energy. Thus light had no mass. Soon a whole spectrum of EM wave frequencies was identified. All frequencies were denied mass or energy.

Clearing up the confusion created an opportunity for Einstein. He gave light and EM radiation the attribute of matter to go along with the wave, thus the confusing duality. He took this forward into the equation E=mc2 which combines the properties of matter and energy into any mass. These individual steps have confused science which should have recognized that anything that is detectible has structure, has mass and can push. More recently the push of light was recognized. There is a band of EM frequencies that corresponds to visible light. Of course the eye can detect light. It does this by having three color cones in the retina which corresponds to red, green, and blue light. The red frequency wave can activate the red cone but not the blue and green cone. In all probability the cones resonate at the given frequency and the intensity is determined by the amplitude of the wave. The frequency selects the color while the amplitude determines the amount of pressure. Likewise gravity as an EM wave should be able to apply pressure to an object and cause it to move.

Beyond just light, in analyzing the expanded spectrum, the push of X-rays was detailed by Compton in 1923 [9]. He aimed X-rays at a plate in his study of the photoelectric effect. The Compton Effect shows that X-ray beams striking a graphite block act like a billiard ball collision with free electrons. The incident wave causes electrons to oscillate. The incident photon transfers some of its energy to the electron. The 'recoil' photons emerging constitute scattered radiation.

In developing this paper the distinction of mass vs. waves led us first down trails of pushing particles and then to pushing waves for more clarity. The two push concepts become equivalent here. The source of gravitational 'attraction' is correctly the result of the push of EM wave particles. Giving gravity this structure suggests many clarifications for the study of physics.

v. **Reduction**

Existing 'attraction' gravity theory provides no details of how mass reveals gravity effects. There is no analysis of what features of mass provide attraction other than an overall mass in large quantities. Is the pull somehow related to atoms or to the nuclei? Likewise, how do those forces radiate out and reverse direction in order to pull back matter bodies?

The ability to have our pushing gravity cause pressure is related to the reduction of amplitude and frequency, and thus of pressure during penetration. It has been observed that less beam amplitude arrives at earth from the sun than from other directions. That leads to the LeSage shadow concept and the 'net gravity' centripetal force attracting earth toward the sun. The beams from the sun that penetrate have less amplitude than the beams from the stars. The 'net' action is more pressure downward and thus that which attracts orbitals is the same 'attraction' gravity we incur on earth.

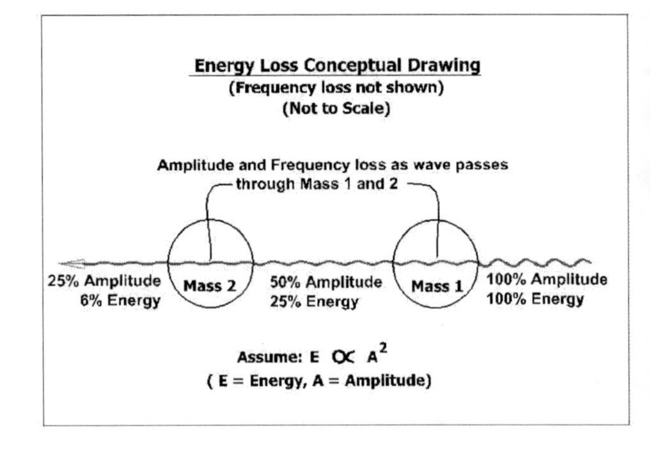

Figure 3 – Gravity Waves Penetrating Masses

Figure 3 shows the wave of amplitude "A" coming from the right and passing through Mass 1 and Mass 2. As it passes through Mass 1 it applies pressure to Mass 1 and its amplitude is reduced by 50%. The amplitude leaving Mass 1 has a value of A/2. Frequency reduction is not shown in the diagram.

7. Inertia

Inertia is the centerpiece of Newtonian dynamics and assumes existing motion of an object is constant until an external force is applied. This inertia is not valid within the EM theory of gravity. EM waves are in equilibrium and constantly interacting with an object. EM waves push the object in all directions which results in stasis (status quo). At any instant of time, one of these waves may cause the object to move but is quickly offset. Over a longer period of time these interactions balance out providing no change in the stasis.

The issue becomes cloudy when the object is in motion relative to the local equilibrium. If the object were in motion then the stasis of spatial pressures means there would be more pressure in the forward direction than in the rear. This is called drag. The object needs help to continue moving in a straight line at a constant velocity. Per Newton's description of inertia the motion can continue in absence of any friction or drag. But space is not void. And some drag/friction automatically occurs within the realm of pushing radiation as motion causes more push to arrive at the front of the motion vs at the rear.

8. Gravity

When two objects are near each other or touching, the reduction of gravitational pressure occurring within each body means beams exit toward each other with lower energy. This produces a low pressure between the objects providing a net push towards each other. This is the mechanism for 'attraction' gravity.

9. Amplitude and Frequency

AM and FM radio waves are good examples of the relation between amplitude and frequency. Both of these EM signals lose amplitude as they pass through objects (the atmosphere and walls). The loss of amplitude indicates an interaction with the atmosphere. The wave would also be putting pressure on the objects and cause them to move. This is the type of interaction by which gravity could cause pressure.

10. Where External Gravity Applies its Pressure

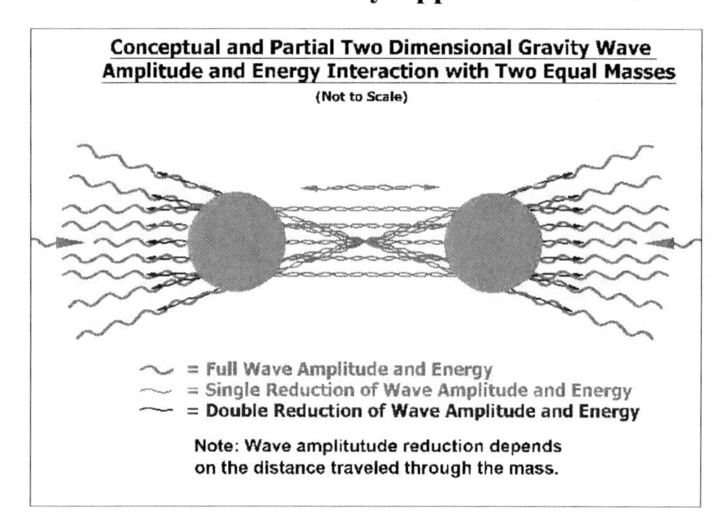

Figure 5 – Attraction Caused by Pushing Waves

All EM waves provide some gravitational type of push. The premise is that gravity beams tend to be very long waves. Shorter waves like light mainly apply pressure to the mass surface. Both long and short waves push in some manner. Light from the sun or stars seems like the antithesis of pressure but light does carry a minimal gravity pressure. Long waves enter and continue to apply pressure throughout their passage through the mass. The longer the distance, the longer the period of penetration for long wave beams, thus the greater the sum of pressures applied to the atomic fields of the mass.

We use and distinguish rays from beams to clarify direction of travel. Rays are emitted transmissions. A beam is similar but carries waves flowing towards matter. Given a range of frequencies, the longer wave lengths penetrate and interact deeper within the matter. Penetrating beams provide pressures over time rather than instantaneously like the surface impact of short waves. The time of penetration is a factor in summing up gravity effects.

11. Predictions

One of the interesting points of reduction is that it cannot be more than 100 percent. Once it gets to 100 percent, the force of gravity is gone, even if there is more mass. In Newton's case,, the more mass there is , there is more force. This implies that there is a maximum amount of gravitational force and a maximum amount of gravitational acceleration.

Dark Matter is proposed because the velocity of the stars at the edge of the galaxy is higher than Newtonian physics can predict. But with reduction, the force of gravity is reduced as it passes through the objects that are in the galactic plane. When it gets to the opposite edge, the force of gravity outward is less while the force of gravity inward is not. This causes a higher net force of gravity towards the center of the galaxy than Newtonian physics. This could explain the high velocity without using dark matter.

12. Conclusions

The purpose of this paper is to investigate that EM radiation has the properties needed to cause pushing gravity.

a. The source of the EM radiation comes from the distant stars.
b. EM radiation is the systematic movement of the EM fabric.
c. The speed of the EM wave is similar to the speed of light.
d. The EM wave will penetrate and pass through objects if the frequency is low.
e. The EM waves will interact and push the object as it penetrates it.
f. When the EM wave interacts with the object, the energy of the wave reduces.
g. The reduced energy of the exiting EM waves cause a lower pressure on nearby objects that allows the pushing waves to move objects toward each other.

We have offered a pushing theory of gravity rather than a pulling theory. This is a 180 degree change in the direction of gravity which could have implications for other areas of physics. We first recognize that gravity exists throughout the universe along with radiation and matter and that the three interact with and penetrate each other. We conclude that gravity is a form of EM radiation and that all EM radiation traveling at speed c can act on, penetrate to some degree, and pressure matter. Meanwhile matter modifies the radiations during their penetration. The strength of radiation, and thus it's pushing potential, is a function of its amplitude while any significant degree of penetration requires long wave radiation beams which exhibit straightness. Penetrating waves lose energy causing an unbalanced push at exit point producing a 'net' downward push. The downward push diminishes and disappears at points distant from the mass as such points are pressured more equally in all 3 dimensions by undiminished beams.

Flaws in the LeSage particle model are overcome here as radiation interaction produces less heat than particles would and rotations of matter can be shown to overcome aberration and drag issues. Although the concepts used to discuss certain actions differ, the model proposed here has no underlying conflict with knowledge learned by applying Newtonian physics. In fact, the concept of pushing gravity overrides the need for several current concepts.

This proposed model does extend knowledge as it suggests that with gravity having an offsetting diminishment factor, there is an absolute limit for gravity forces which negates the unlimited measure of attraction gravity theory, that lead to black hole and expansion concepts.

A potential value of our work is to redefine forces so they become a mechanism to the orbiting and curvature to central body forces of gravity. Kepler's laws suggest that relationship.

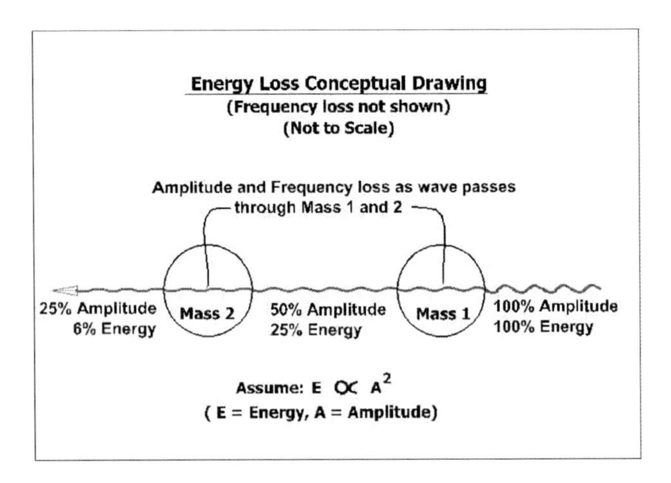

References

[1] - I. Bernard Cohen, *Introduction to Newtons Principia,* Harvard 1978, page 177. Newton stated that "Nicolas Fatio de Duillier found the only possible mechanical cause of gravity"

[2] - Matthew Edwards, editor, Pushing Gravity; Apeiron 2002, page 10, Le Sage's Theory of Gravity http://en.wikipedia.org/wiki/Le_Sage%27s_theory_of_gravitation,

[3] - Shortley and Williams editors, Elements of Physics, Prentice Hall 1955, page 95, Newton's law of Gravitation http://www.relativitycalculator.com/Newton_Universal_Gravity_Law.shtml

[4] - Halliday & Resnick, Physics, Wiley & Sons. Combined Edition 1966, Section 48.1, Waves and Particles, pg 1203.

[5] - Sound Waves – Intensity of sound is proportional to the square of the amplitude. http://physics.info/intensity/

[6] - Halliday & Resnick, Physics, Wiley & Sons, 3rd edition. 1978, Photoelectric effect, section 49.4

[7] - Energy and frequency of EM waves - hv = E0 + Kmax http://en.wikipedia.org/wiki/Electromagnetic_radiation

[8] - Radiation pressure, http://en.wikipedia.org/wiki/Radiation_pressure, http://www.pa.msu.edu/~schwier/courses/2014SpringPhy184/lecture37.pdf

[9] - Halliday & Resnick, Physics, Wiley & Sons, 3rd edition. 1978 Compton Effect, section 49.6

The Gyroscope

Robert de Hilster
23344 Carolwood Ln #6409, Boca Raton FL 33428
e-mail: robert@dehilster.com

Gyroscopes behave in very strange ways. Many demonstrations are given with little explanation just to keep the audience guessing. When you search for the answers, you are told that it is the angular momentum, the right hand rule and you are shown some equations. But the explanation is not very clear. This paper suggests that angular momentum is not the best explanation. Rather, linear velocity, cohesion, and gravity are all that is needed to understand the strange movements of the gyroscope.

1. Gyroscopic Motion

Take a toy gyroscope, set it on a pedestal and it falls down. Oops! I forgot to spin it. OK, take a toy gyroscope, spin it, and set it on a pedestal. It does not fall. It moves around and around and then after a short period of time, it falls.

What keeps it up?

What causes it to move in a circle?

Is it weightless??

It seems quite clear that this strange motion of gyroscope is caused by the spinning of the rotor.

2. The Normal Explanation

When you read reports that explain the gyroscope, here is what you find:

Angular momentum

Right hand rule

Math

Using these ideas, showing drawings of the gyroscope, and showing the force vectors; we are expected to understand how it works.

But it does not help!

Is it the Momentum?

There are two types of momentum, linear and angular. We have all experienced both of them. When you were young, were you hit by a baseball? Did it hurt? Of course it did. The faster the ball flies or a ball with a greater mass, it will hurt more because it has more momentum. That's linear momentum.

Have you ever tried to stop a spinning bicycle tire? The tire will push your hand in the direction of its spin. Depending on the mass of the tire and its angular velocity, the angular momentum can be very strong. It will almost tear your fingers off.

Unfortunately, the common explanation states that the direction of angular momentum **L** is determines by the right hand rule.

Is it the Right Hand Rule?

Figure 1 is an example of the right hand rule. The fingers of the right hand are curved in the same direction as the spin of the rotor. The thumb determines the direction of the angular momentum. This direction of the angular momentum is perpendicular to the spin of the rotor, while the direction of linear momentum is in line with the linear velocity. Can you explain that?

Is the **ω** by the thumb is wrong? Should it be **L** instead?

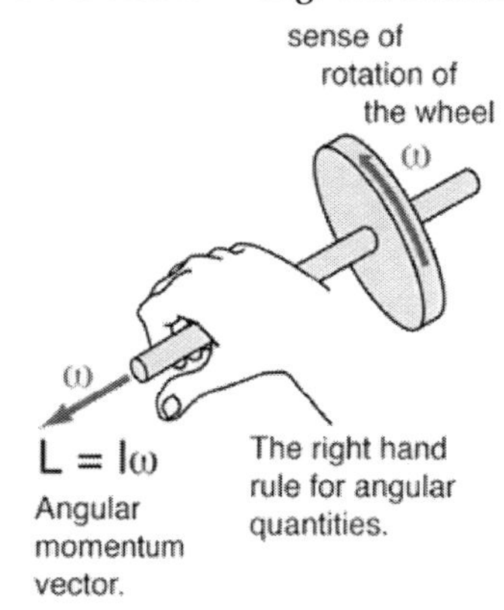

Figure 1 - The Right hand Rule

If the spinning tire pushes your hand in the direction of the spin, why is the angular momentum of the gyroscope described as being perpendicular to the spin?

Is there a feature of the gyroscope that requires this to be true?

Is it the Math?

$$\vec{L} = I\vec{\omega}. \qquad (1)$$

Equation 1 is the equation for angular momentum. It indicates that the angular momentum **L** is a vector that should be in the same direction as the angular velocity **ω**.

$$I = {}^{2}/_{5}\, mR^2 \qquad (2)$$

Equation 2 is the moment of inertia of a solid sphere. It is generally understood to be equivalent to linear mass as used in the equation for linear momentum. I is stated in units of Kg m/s.

$$k = \sqrt{I/m} \qquad (3)$$

Equation 3 is the radius of gyration. Concentrate all of the mass of the rotor in a particle. 'k' is the distance that the total mass of the rotor is from the axle. With the mass in this position and spinning at an angular velocity of ω, the value of angular momentum is maintained.

Figure 2 shows a typical position where the mass could be concentrated and is moving at the angular velocity of ω. Clearly the effect of angular momentum shown in Figure 2 has the same effect as the spinning tire.

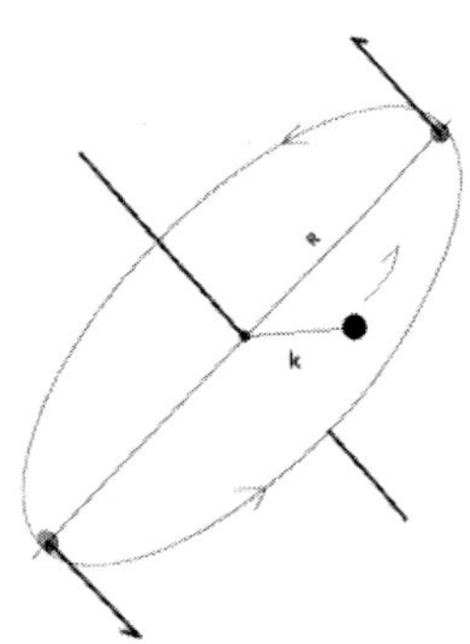

Figure 2 – Mass Concentrated at Distance k

Caused by Cross Products?

Torque is normally explained using the cross product of a force vector and a distance vector. It turns out that the resulting vector has a magnitude and direction. For some reason the resultant vector is defined as being perpendicular to the force and distance vectors. Because it is the only choice left? But to emphasize the confusion, the resultant vector is called a pseudo-vector. It's not real!

I have difficulty with the idea that distance is a vector. So the resultant torque vector is a product of force times the distance, and the direction is in the same direction as the force vector, that works for me! So the angular momentum vector is in the same direction as the angular velocity vector, which is what we experience with a spinning tire.

3. More about Angular Momentum

It is an Effect

Angular momentum is not a cause, it is an effect. It is a combination of linear momentum and cohesion. So, what is the cause of linear momentum? Even Newton did not answer this question. But he did state the idea in his first principle of motion. He called it inertia. I prefer to call it linear momentum or linear velocity. In any case it happens and we just have to accept it until we find a better answer.

Cohesion is my word that combines all the forces that hold particles together. They are nuclear forces, electromagnetic forces, and even gravity. It is this force that keeps a given particle at a given distance from the center of rotation.

There is no outward component

One effect of spinning is the existence of the equatorial bulge. No matter how you look at angular momentum, there is no way that it can produce an outward radial component. It moves in a circle.

Linear momentum has a radial outward component and as such can explain the equatorial bulge.

4. Stability and Position

The chart below compares the values of momentum and velocity using a lead ball and the earth. The lead ball has a radius of 0.1 meters and is spinning at the rate of 6,000 RPM. The earth has a radius of 6,371,000 meters and spins at 1 revolution per day.

	Lead Ball	Earth
Angular Momentum	1.19E+2	7.05E+33
Linear Momentum	2.98	2.76E+24
Angular Velocity	6.28	7.27E-7
Linear Velocity	62	463

Chart 1 - Comparison

When observing the motion of a toy gyroscope, it is clear that it does not take very long for the gyroscope to fall. On the other hand, the earth has been spinning and orbiting for a very long time and has not fallen into the sun. The momentum of the earth is much greater than the momentum of the lead ball. It seems that momentum speaks to stability.

However when calculating the next position of the rotor, velocity is used. So, for the rest of the paper it is not the angular or linear momentum that is used, it is the linear velocity.

5. The Next Position

Using an ideal gyroscope (the mass is perfectly balanced) with linear velocity and cohesion, a gyroscope will maintain its position. See Figure 3.

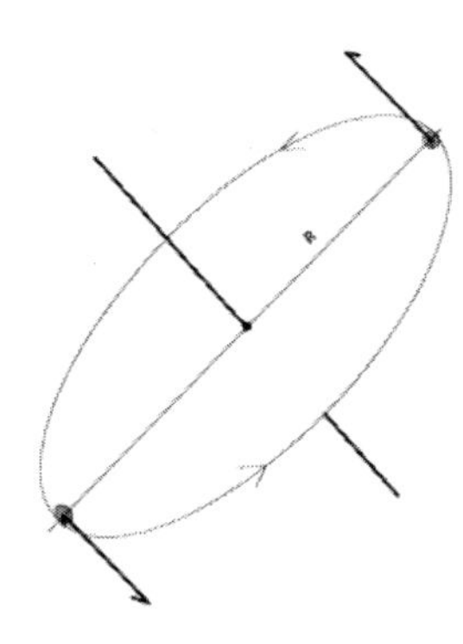

Figure 3 – Linear Velocity and Cohesion

There are two small pieces of the rotor shown in Figure 3. Each one has a linear velocity as shown. If they were free particles, they would fly out in the direction shown. Of course they are not free because they are held in by cohesion

So, we can use the model to predict where the two pieces will be in a small increment of time. It is not only the direction of the linear velocity nor the direction of the cohesive force. It is their magnitude. The linear velocity of the two small pieces is shown in Chart 1. But it is the cohesion that is the strongest. It is so strong that the radius R of the rotor shows no visible signs of change.

To calculate the next position, assume the small sphere is free and would move a distance Δx, Δy, and Δz in the time of Δt. Then use cohesion to move it back to a fixed distance R. With no other forces present, each small piece will move in a circle around the axle.

It will continue to do this as long as there is no imbalance of mass or forces.

6. Perfect Balance

Figure 4 shows a perfectly balanced gyroscope.

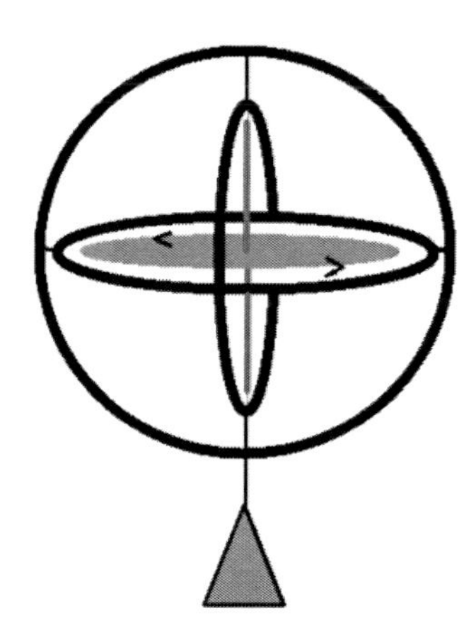

Figure 4 – Perfectly Balanced

If the mass of the gyroscope, including rotor and gimbals, is perfectly balanced; and the axle is perfectly lined up with gravity; and there are no other forces, the gyroscope will hold its position for ever. But the real world is not so kind. Any unbalance of mass or alignment or external forces will cause the gyroscope to tilt and eventually fall.

7. Falling Over

So let's assume you have a spinning gyroscope, you set it on a pedestal, and the axle is perfectly aligned with gravity. If there is a very small imbalance of mass, gravity will act on the imbalance and will tilt the axle a small amount in the direction of the extra mass. The amount of tilt that the axis assumes depends on the amount of the imbalance of mass. This explains why the gyroscope initially drops to a small or larger angle. This type of imbalance will cause the gyroscope to wobble; not to precess.

It turns out that while the gyroscope is precessing it continues to fall or to have a greater tilt. This happens as the linear velocity of the rotor slows down and force of gravity becomes more dominate.

8. Precession

Once the axle is tilted, the gyroscope will precess. Precession is the circular movement of the axle. We observe the movement of the axle, but it is the rotor that is in control. Further, the precession appears to be consistent in direction and velocity.

Direction

If you spin the rotor clockwise (CW), the axle moves CW. If you spin the rotor Counter-clockwise (CCW), the axle moves CCW. It happens every time. The direction of spin is always viewed from above the gyroscope and in the direction of the force of gravity.

We say the earth spins CCW because we observe the spin from the North Pole. But the reality is that the North Pole is established when we observe the CCW spin. So the same convention is used for the gyroscope.

The direction of spin is deterministic because the environment on your desk top has almost no extraneous forces that could change the motion. There is not much wind, nor magnetic and electrostatic fields, nor mechanical forces. But there is gravity. And for the direction to be consistent, the imbalance of gravity must be in the same direction all the time. Figure 5 shows a tilted gyroscope spinning CCW with the force of gravity applied to two pieces of the rotor.

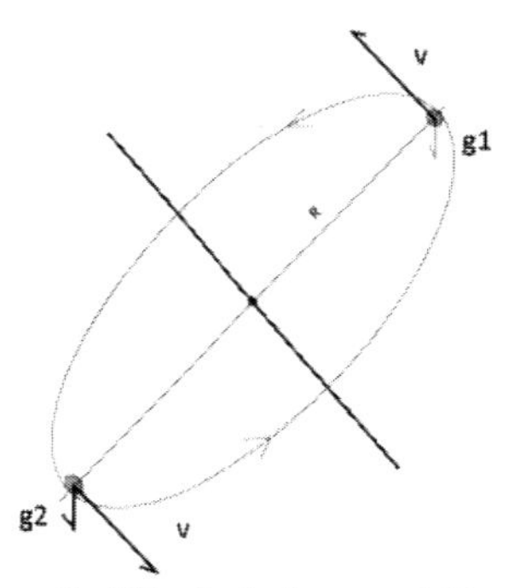

Figure 5 - The imbalance of Gravity

In the figure, the gravitational acceleration for the lower piece (g2) is greater than the gravitational acceleration of the upper piece (g1). Although both pieces are being pushed down the there is a net torque on the rotor at position g2. The common explanation of precession uses torque but it is applied at the center of the rotor and no further explanation is given.

The purpose of putting the net torque on g2 is to calculate the next position of the lowest point of the rotor. The lowest point of the rotor is marked by the position of the top of the axle. By calculating its next position, the movement of the rotor can be determined. The rotation of the axle will follow.

This lowest point moves slowly around in one direction. All the points below the horizon have a net force pushing them down. So as the linear velocity moves the points CCW, and the cohesion holds the points to a fixed distance from the axle, gravity pushes the points down. And so the lowest point on the rotor moves CCW.

Although this explanation seems reasonable, I am not absolutely convinced. The calculations need to be done.

Velocity

There is a video of Professor Laithwaite demonstrating the mysteries of the gyroscope to a young group of students in 1974. Reference [1] provides the link to this video. In the 4th part of the video, he demonstrates that the precession rate is determined by the amount of torque applied. There is a derivation of the velocity of the precession provided by Georgia State University [2]. The final result is shown in Equation 4.

$$\omega_p = mgd/I\omega_s \quad (4)$$

ωp is the angular velocity of precession
m is the total mass of the gyroscope
d is the distance
from the pivot point to the center of mass
I is the moment of inertia
ω_s is the angular velocity of the rotor

The torque in the equation is mgd, so it is clear that if the torque is doubled, the velocity is doubled. But I do not agree with how it was done. It starts with the angular momentum being directed along the axis of the gyroscope. If it starts with this, I can't proceed. I must go back to Figure 5 as my basic model and develop it further. It would get more complicated and be difficult to calculate, but it would be based on what I consider to be valid physics principles.

However, equation 4 may still be correct since I have not run a test or performed any calculations.

9. Calculating the Direction and Velocity

Figure 6 is my model for calculating the direction and velocity of a spinning gyroscope.

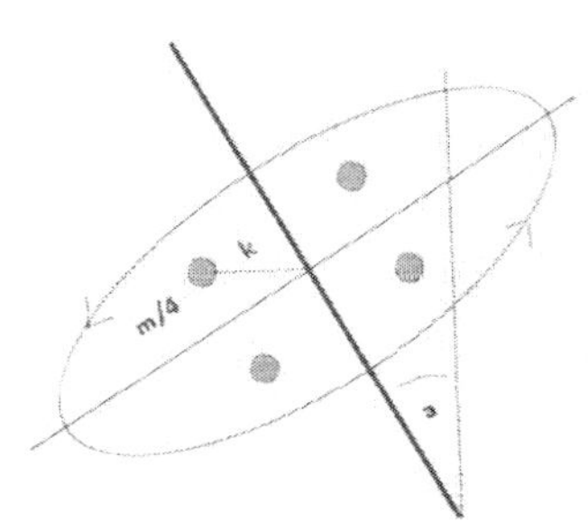

Figure 6 – Model for Calculation

The mass of the rotor is broken into four equal parts and placed at equal intervals around the axis each at a distance of k The angular velocity is **ω** and the angle of inclination is **a.** The pivot point of the axle is at sea level and is the origin of the c0-ordinate system. The forces of gravity are applied at each of the four points. As a starting point for the calculation, I will use this model and the following algorithm.

The lowest point of the rotor is in the same direction as the direction of the axle.

Calculate the next position of the rotor for at least one revolution or more.
At the end, find the lowest point of the rotor.
The rate of precession is the change in angle of the lowest point divided by the elapsed time.
The direction of the precession is determined by the sign of the change in angle.

Compare the result with the GSU equation.

The GSU equation does not emphasize the direction.

10. Summary

Weightless

Clearly the spinning gyroscope appears to float. But is it weightless? That is, does it have zero weight? Equation 4 has the term 'mg' that represents the force on the center of the gyroscope and 'm' is the total mass of the gyroscope. The equation is stating that all of the mass is being acted upon by gravity. So the mass is not less and gravity is not less. So it is not weightless!

It floats because the linear velocity is greater than the force of gravity. So the rotor tries to stay where it is. As the rotor slows down the linear velocity decreases and gravity has a greater effect causing the gyroscope slowly fall over.

Angular Momentum

Angular momentum is used to explain the strange movements of the gyroscope. However, it is called a pseudo-vector because it is not real, which makes the right hand rule null and void. Angular momentum does speak to the issue of stability. It is linear velocity that determines the next position of the rotor and provides the best explanation of the gyroscopes strange behavior.

Precession

Precession does seem deterministic in both direction and speed.

References

[1] - Professor Eric Laithewaite 1974 Demonstrations
http://www.intalek.com/Index/Projects/Research/Laithwaite/Laithwaite1974.htm

[2] - Georgia State Universities equation for angular momentum
http://hyperphysics.phy-astr.gsu.edu/hbase/top.html

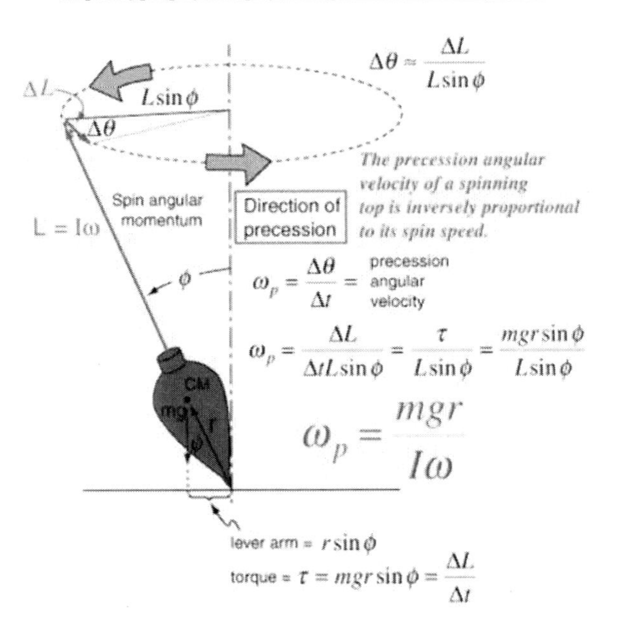

Bertrand Russell and "CONTINUITY"

Peter F. Erickson
P.O. Box 87667, Vancouver, WA 98687
e-mail: peterfoserickson@aol.com

Central to establishment's concept of continuity is that there be no "next-to." It is integral to set theory and modern mathematics. Bertrand Russell noticed that it is in conflict with the common sense understanding of differential equations. Nonetheless, he accepted it. This led to a bizarre notion of a "physical object." Is there an alternative to this concept of continuity? Yes, there is.

1. Introduction

1.1 Who He Was

Bertrand Russell (1872-1970) was among the makers of the modern intellectual culture. He was a major developer of modern mathematical logic. His work has affected not only the character of Anglo-American academia, but also the pathways to what became "computer science." In 1950, he won the Nobel Prize for Literature.; (the Noble Prize for Mathematics does not exist). He was one of the most influential intellectual commentators of the general issues of his time. Einstein said: "I owe innumerable happy hours to the reading of Russell's works, something I cannot say of any other contemporary scientific writer, with the exception of Thorstein Veblen."

1.2 The Modern Theory of Continuity

According to this idea, the points on a line are never next to each other, but that between any two points, another can be inserted. Lord Russell endorsed it completely. He proclaimed that the "physical continuum . . . the continuum assumed in physics; that is precisely the same as the mathematical continuum, since it is assumed that the real numbers are necessary and sufficient for the assignment coordinates.."

In The Analysis of Matter, Russell wrote that "there is absolutely no apriori reason why everything should go by differential equations, since, even then, causation does not really go from next to next; in a continuum there is no 'next'. It is, at bottom, because 'next-to-next' seems natural that we like a procedure of differential equations; but the two are logically incompatible, and our preference for the second on account of the first proceeds only from logical confusion."

Here, the common sense notion that there must be a point with no breach in between it and another point is denied. Or, as Russell put it elsewhere in the same work, "However the mathematical treatment may confine itself to infinitesimals, relations between points whose distances are finite must be presupposed if the infinitesimal calculus is to be applicable. " His meaning is clear. There can never be any differential change in an interval between two points lying below the level of the finite. Neither can there be a differential change in the interval between any sub-finite point and the nearest finite terminal, since there cannot under that supposition be a next-to. Nor can an irrational distance between points in the calculus interval exist. Only the finite rational is possible.

1.3 The Plan of this Paper

In this article, the consequences of Russell's acceptance of the modern idea of continuity is traced through his theory of irrational numbers, then through his conception of a point, through his notion of logic, and last, through his idea of the empirical object. After which, an alternative to the modern theory will be suggested.

2. Continuity and Irrational Numbers

2.1 Background

For the classical Greeks, a number could best be represented as a length, the number 2 being twice as long as 1; the number ½ , half that of 1, and so forth. The historical number line is based upon that understanding. Each extension of the line from point to point establishes a greater number than any of its predecessors. An irrational number, such as □2, cannot be stated as any kind of an aliquot fraction: it is incapable of being located exactly, regardless of the base used, whether that of 10ths, 12ths, or whatever. It can only be approximated by rational numbers. Yet, it is a specific length and therefore must exist. As I have written in a former work: " . . . our digital approximation, even if we were able to carry it beyond the ability of science to find it, would actually be something different from □2. The difficulty is not practical, but rooted in the nature of reality."

2.2 Russell and Dedekind

Russell attempted to reduce this magnificent fact to a mere manipulation with symbols. He brought up the notion of the "cut" which had been introduced by Richard Dedekind in the 19th century. According to this notion, □2 is established by a division between the series of rational numbers below it and the rational series above it. Although Dedekind 's account of this was incoherent, his notion was reminiscent of the classical idea of a point lying between the two series, this point being the terminus of the length of an irrational number.

But Russell wanted to eliminate any spatial references.

2.3 Russell's Wrongful Analysis

Russell shifted from single numbers and points to ratios between numbers. Doing this is completely in consonance with his beliefs about continuity: that there is no obvious next-to with respect to the ratios between two numbers; that given any two ratios, another could always be inserted. (This tactic allowed him to obviate an objection which was sure to have been brought forth against any attempt to disqualify irrational numbers on the grounds that they cannot be precisely set down, this objection being that the largest rational number less than □2 cannot be precisely found either. Should Russell have been required to increase the sum in order to obtain a closer approximation, he needed only insert a different ratio between two lesser ratios already present. That way, he did not have to concern himself with the point just next to □2.)

Having done that, he argued: "From the habit of being influenced by spatial imagination, people have supposed that series must have limits in cases where it seems odd if they do not. Thus, perceiving that there was no rational limit to the ratios whose square is less than 2, they allowed themselves to 'postulate' an irrational limit, which would fulfill the Dedekind gap. Dedekind set up the axiom that the gap must always be filled, i.e., that every section must have a boundary."

Note here that the term, "boundary," is spatial. But let us suppose for the sake of the argument that rational numbers can be imagined without any reference to a number line.

With this non-spatial "boundary" as the key concept, he then trivialized Dedekind and, even more, the number line idea which rendered the "cut" intelligible.

Russell declared that there are series which have a boundary and those that do not.

"A series is called "Dedekindian' when every section has a boundary, upper or lower as the case may be." Russell argued that because of their "spatial imagination," people figured that since there was no "rational limit to the ratios whose square is less than 2, they allowed themselves to 'postulate an irrational limit, which was to fill the Dedekind gap. .Dedekind . . . set up the axiom that the gap must always be filled, i.e., that every section must have a boundary. It is for this reason that series where his axiom is verified are called 'Dedekindian.' But there are an infinite number of series for which it is not verified.

"The method of 'postulating' what we want has many advantages; they are the same as the advantages of theft over honest toil. Let us leave them to others and proceed with our honest toil."

Then began his toil: "Let us confine ourselves to cuts in which the lower section has no maximum; in this case we call the lower section a 'segment.' . . . . A 'real number' is a segment of the series of order of magnitude. An 'irrational number' is a segment of the series of ratios which has no boundary. A 'rational' real number is a segment of the series of numbers which has a boundary."

The expression of a rational number either terminates or repeats itself in patterns. If 10 is used as a base, for example, digital fractions representing 1/3 and 1/7 are incapable of finite completion; they can only be stated with recurring groups of numbers. Russell's irrational number, apparently, is supposed to have no boundary, because, not only is it not complete-able, but it has no repeated patterns either.

Has he succeeded in depriving "boundary" and "segment" of all spatial reference by then keying them to "order of magnitude"? Consider the length of the hypotenuse of a right isosceles triangle one unit in length on each side – probably the origin of the discovery of the irrational number. According to the Pythagorean theorem (as well as common sense) that hypotenuse must have a length. Where it joins two unit sides of the triangle are its boundaries. It cannot be any longer or shorter than that. The difference between the length and the nearest rational length below it must exist. If this is to be disposed of on the grounds that it is spatial, then so must the measurement of the hypotenuse of triangles with rational sides like 3, 4, and 5. And with the Pythagorean theorem, there goes Russell's blessed science of physics.

But let us see what he comes up with. " . . . □ 2 is the upper limit of all those segments of the series of ratios that correspond to ratios whose square is less than 2. More simply still, □2 is the segment consisting of all those ratios whose square is less than 2."

His "limit" is less than the square root of two. But that can only be an approximation. It is not and cannot be equal to that root. In order to be the same as the number whose square is equal to 2, there would have to be something beyond these fractions. He realized that he could not use a continuous sequence of rational numbers, because beyond the largest single rational number, there would always be □2 and perhaps even one or more unrecognized irrational magnitudes just before that. But that he cannot have. The tactic of switching from numbers to ratios of numbers was also amiss, since, by definition, an irrational number cannot be a ratio or a series of ratios.

His goal was to deprive the irrational number of any objective status, to offer no-boundary as if it were an adequate account of irrational magnitudes. With the number line, an irrational number exists, even though not completely locate-able. But it has to have a specific identity. Yet, he can accept only the rational. Since no one would believe him if he were to declare openly that it does not exist, he said it has no boundary; the semblance of solidity is offered in lieu of substance. That way he did not have to recognize an irrational number as an existent, albeit incapable of being perfectly known.

In this regard, he was like so many rationalistic types who think it a scandal that they cannot know everything, at least in principle. Russell's calling the acceptance of the factual nature of irrational numbers "dishonest," was not so much a tactic to disarm his opponents through a phony objection as it was a kind of religious commitment -- what he darkly referred to as "honest toil." Although Russell was a proclaimed atheist and situationist in ethics, he had a deep intention.

How is it that he accepted this contradiction? It is because he embraced a notion of continuity which allowed no next-to -- which was full of holes. This can be seen in his defining a rational number as a segment of the series of numbers that has a "boundary." This leaves unasked the question, what are they ratios of? If one considers only the fractions below 1 and leaves out the irrational numbers below it, the rational fractions together cannot together equal a unit. With no next-to, there is vacuity before and after every point. The unit would be shrunk. Therefore, he could accept a hiatus in place of the actual □2.

And, since for Russell, an irrational number has no substance, he is stuck with only rational changes of velocity in the second derivative in the differential calculus. The possibility of irrational derivatives is excluded by philosophical fiat -- what he had received from Georg Cantor.

3. Russell's Conception of the Point.

Let us begin with his conclusion. "A point is a co-punctual group which cannot be enlarged without ceasing to be co-punctual. " Note here that a single point is a group. This already does away completely with Euclid's definition of a point as that which has no parts. What then are the parts of the group? Sub-points, I suppose.

His "point" has little to do with a position in space. It is to be topologically defined. - the order between the parts of his point is to be considered. It can move about, but its parts are to retain their order.

What then is it (or is it them)? He gave an example. He asked that we imagine "a three dimensional "Euclidean numerical space, i.e., the manifold of all ordered triads of real numbers (x, y, z) with the usual definition of distance. " He didn't quite mean Platonic numbers separated by a "distance." Instead, he wrote: "Consider, in this space, all the spheres having a given radius and having centers whose coordinates are rational." No irrational radii, of course. Nor are these centers points in the classical manner. "Let us define a group of these spheres as 'co-punctual' if it is such that every four chosen out of the group have a common region; and let us define a co-punctual group as 'punctual' if it cannot be enlarged without ceasing to be co-punctual."

His "point" is nested by its surroundings. Where or when does nearness not belong? Suppose that a sphere not concentric to the spheres centering about some region were added; either that, or suppose that that a sphere were added that intersected only one of these inner spheres but did not surround the inner core. In either case, the collection would no longer be co-punctual.

But this was to be only a three-dimensional prologue. After some discussion, Russell's co-punctuality comes to a five term relationship that, he added, could only work in a four dimensional manifold. His "point" is only the innermost part of some kind of a nest. It hardly bears even a cousinly resemblance to the classical understanding. It is only relatively the smallest. And, of course, no next-to. "We will call two points 'connected' when there is an event which is a member of both."

This brings up a question: Since, for Russell, a point is just what is inside a final enclosure as determined by its context, it follows that the size of a "point" will vary with context. Given this, consider what might happen if points of different sizes were positioned so that from the observer's standpoint, one of them was placed above the other one (although not touching it). Let the closer Russellian point be labeled A and the more distant one, B. Would it not be possible that the size of point A might be large enough to cover the both point B and also the area comprised by point B′ near by but not next to point B? If this is possible, then why cannot there be a point B″ between B and B′? Since A covers more than B _ B′, why cannot the latter be completed next-to-next. But Russell postulated that B _ B′ cannot be completely connected. It is clear why Russell had to resort to topology which considers the order of points in abstraction from the distances and areas between them.

Topology has its own problems. How to select the pertinent order? If the selection of the points is based on nearness, then distance cannot be excluded. If distance is avoided, then different transpositions of points must be allowed. Topologists end up soberly describing impossibilities like Klein bottles. Suffice it to say that Russell himself saw problems with it. He once whimsically suggested: "If one could imagine intelligent beings living on the sun, where everything is gaseous, they would presumably have no notion of number, any more than of 'things.' They might have mathematics, but the most elementary branch would be topology. Some solar Einstein might invent arithmetic, and imagine a world to which it would be applicable, but the subject would be considered too difficult for schoolboys.".

4. Russell's Logic: Some Considerations

4.1 Russell's Idea of Implication

Having grasped the fundamental inability of Russell's thought to come to terms with irrational numbers and his situational account of the point, let us consider his argument on formal implication. Since there is no spacial next-to in modern mathematics, it should be of little surprise that its leading logician would fashion a system in which necessity is hard to find.

For Russell's Principia, there is an indispensable deduction, the one called "implication, "signified by " □. " Almost everybody is familiar with his Truth table. Here, the most basic form, the four place, is presented in a non-tabular manner. (A) p is True □ q is True; (B) p is False □ q is False; (C) p is True □ q is False; (D) p is False □ q is True.

This implication is simply a disjunction between (A), (B), and (D). p will imply q whenever both are T; both are F; or p is F but q is T. In other words, p implies q, except (C), where p is T, but q is F.

4.2 Some Difficulties With it

Admirers of the new logic contend that this new kind of implication is superior to the idea of implication found in the older logic books. Those books state, in effect, that "if p is true, and it is true that p implies q, then q must be true." This, the advocates of the new logic argue is completely satisfied by (A). Moreover, (C) "claims that if p is true and q is false, then it is false to say that p implies q. " "Hence, " (A) and (C) "are consistent with the old idea of implication, and therefore the new 'material implication' includes the old idea. "

But is the new idea better because it includes the old idea and is, therefore, more general? Here are two kinds of generality: The first kind consists of an inventory which includes every item in the list - in this case, (A), (B), (C), & (D). A concrete example would be a list of all shoes, comprising the freshly made, the serviceable, and the broken down. A second kind of generality would include that which they have in common. But this can come with a diminished meaning. As an extreme example: if a bottle of an elixir were tossed into a sewer, the resulting combination could be generalized in terms of that which is the same throughout.

From the first days, critics have pointed out difficulties with Russell's theory of implication, most notably that it is a logic without necessity. Russell's "material implication" results between p and q under these three conditions: (i) T—T, (ii)F - F, and (iii) F—T. Consider the following examples (i) Caesar was stabbed; Trotsky was stabbed. Both propositions are true. Caesar's being stabbed implies that Trotsky was also stabbed. (ii) Water is heavier than lead; Ayn Rand wrote Gone With The Wind. Both propositions are false. That water is heavier than lead implies that Ayn Rand wrote Gone With The Wind. (iii) Caesar was defeated by Alexander the Great at the Battle of the Little Bighorn.; 2 + 5 = 7. The first is false; the second one is true. Therefore, the first implies the second. In all three cases, there is implication without necessity.

If p is false, not only will it imply the particular false proposition named, but any other falsehood. The false proposition that Caesar was defeated by Alexander the Great at the Battle of the Bighorn implies that Alexander the Great was defeated by Caesar at the Battle of the Little Bighorn. Moreover, since of any pair, p and q, if the second is true, it is implied by the first. That FDR was elected four times is implied by Cyprus is an island; that red is a color is implied by 2 + 2 = 5.

Brand Blandshard shows more: "The same holds true of other essential ideas of logic, such as necessary, possible, and impossible. These two are defined as truth functions. Russell says that □x 'will be necessary if the function is always true, possible, if sometimes true, and impossible if never true.'" Blanshard demonstrates the inability of Russell's logic to handle all three. We shall confine our examination to the third case, the impossible.

Let □ stand for "past President of the United States born since 1823." and □ stand for "born in land which is or would be the State of Oregon" Then the statement that it is impossible that any past President of the United States born after 1823 should have been born in the land which is or would be the State of Oregon would be (x).(□ □ □□), or "no U.S. President born after 1823 was born in what is or would be the State of Oregon." It denial would be □ [(x).(□ □ □□)] or "It is not true that no U.S. President born after 1823 was born in what is or would be the State of Oregon." Both the assertion and its denial fail to mean what has been asked for. To say that it is impossible for there to have been a President of that description is to say, not that there has been no such a President, but that there was something in our in the American way of life that would have made it impossible. (Blanshard's own example in his 1962 book was a Black President, which is stronger than the illustration offered.).

Blanshard summarizes: "Actual reasoning appeals to musts and cannots that belong to the content of what is asserted; without these, the nerve of the inference would be gone and validity would be meaningless. It is therefore the business of logic to analyze these connections. Principia does not analyze them; it ignores them. At the crucial points it turns its face away from the necessities of actual thought, and substitutes for them ideas which it regards as more convenient in the elaboration of its system. For entailment it substitutes implication, material and formal, in neither of which is there anything beyond general accompaniment. For inconsistency it substitutes 'not both true,' from which necessity has evaporated. Possibility and impossibility are cut down to assertions of 'some' and 'none.'"

It might be argued that Russell's limping logic was due to the extreme empiricism he obtained as a youth while studying John Stuart Mill's System of Logic. Mill, famously, taught that all science is provisional.

But Russell broke decisively with Mill on the question of logic and algebra. Whatever of Mill's empiricism Russell had kept, these remnants were intensified by his embrace of what he called the "Cantorian continuum."

5. Russell's Conception of the Empirical Object

Russell did not stint from intellectually living within his conviction. This can be seen in his analysis of perception. According to the older idea of physics, there is a difference between the real table of physics and the one that we perceive. This real table underlying the appearances would differ in several ways from what we perceive., but it still would exist as the convergence of the directions from which it was perceived. For Russell, however, "it was mistaken to regard the 'real' table as the common cause of the 'appearances' which the table presents (as we say) to all observers." That "instead of looking for an impartial source, we can secure neutrality by the equal representation of all the parties." In sum, instead of the table being the central object, it is to be regarded as the whole assemblage of appearances founded at the many places occupied by those doing the observing -- "the set of all those particulars which would be called 'aspects' of the table from different points of view."

An "appearance" is not an effect, but is "actually part' of the object itself, in the sense in which a man is part of the human race." In other words, all of them together and the places an observer might have been but is absent constitute the object -- in the same way that a single individual is a member of the whole human race, even if he lives on a desert island and no one else is aware of his existence.

Russell added: "I think that when I see (say) a penny, what I perceive is one member of the system of particulars which is the momentary penny, that it is a member which is situated (according to one meaning of 'situation') in a certain part of my brain. I think that, very near this part of my brain there are closely similar unperceived particulars which are other members of the momentary penny; there is no solution of continuity in passing from what I perceive to the outside particulars dealt with by physics."

Russell's penny includes those perspectives which an observer might have seen had he been present, even if no one was there to look. As Russell said, it "shows that what we call a material object is not itself a substance, but is a system of particulars analogous in their nature to sensations and in fact often including actual sensations among their number. In this way, the stuff of which physical objects is composed is brought into relation with the stuff of which part, at least, of our mental life is composed."

5.1 Lovejoy's Refutation

By now, some readers may be wondering, what has this to do with the nature of continuity and the modern mathematical notion that there might be no next-to. As will shortly be shown, it has EVERYTHING to do with it. This can be gleaned from Arthur O. Lovejoy's great book, The Revolt Against Dualism: Russell " offers . . . a definition of 'the place where (in perspective space) where a thing is'; and this, upon close scrutiny, oddly turns out to be a definition of the place where the thing is not. It is reached as follows. We again consider the penny which we found appearing in many perspectives. 'We formed a straight line of perspectives in which the penny looked circular, and we agreed that those in which it looked larger were to be considered nearer to the penny. We can form another straight line of perspectives in which the penny is seen end-on and looks like a straight line of a certain thickness.' These two lines, 'if continued will meet in a certain place in perspective space, i.e., in a certain perspective, which may be defined as 'the place (in perspective space) where the penny is. ' Put more briefly, this appears to mean that 'the penny' is at the center in the 'one all-embracing space' upon which lines drawn through each of the series of similar (but unequal) 'aspects' of it would converge. But we have seen Mr. Russell elsewhere insisting that the penny is not in any place, but simply the complete set of aspects scattered through space. This seemed to be the great point in the enunciation of the new definition of a physical thing; but it is manifestly inconsistent with the definition of the place 'where a thing is.' At most only a part of the of a thing – i.e., one of the aspects -- could be at that (or any other) place."

Professor Lovejoy continued: "And if we consider the Russellian conception of a 'physical object' yet more closely, we shall see that not even a part of such an object can properly be said to be in the place where (in the definition cited) 'the object' is said to be. This is evident not only from Russell's statements but also from the implications of his premises. It is, in the first place, the consequence, in the terms of this theory, of the empirical fact which common sense would describe by saying that an object (as a whole) cannot be seen or photographed from the place which it occupies. . . . In Mr. Russell's own words: 'As a rule . . . even when the center of a group' of appearances 'is occupied by a percipient, it nevertheless contains no member of the group, not even an ideal member: the eye sees not itself. A group, that is to say, is hollow: when we get sufficiently near its center it ceases to have members.' That same conclusion is implied in Mr. Russell's objections to that 'unnecessary thing in itself,' the 'real table' of common sense, is equally evident. If there were a central aspect to the table, it would be a mere quibbling of words to refuse to call it the table; for it would have all the attributes ascribed by common sense to the definitely located causal object which Mr. Russell has repudiated. It would be a material reality. . . . Mr. Russell plainly intended to deny the existence of something which people generally have believed in; but his denial would have come to nothing if he had left either central objects or (what in all but name, would come to the same thing) central members of groups of aspects, in the places assigned to the 'real' tables, pennies, etc., of common sense. The only adequate reason, in short for redefining the 'physical object' as the diffused collection of all the aspects lies in the assumed disappearance of any reality of the kind at the middle of that collection. All material things, then, in Mr. Russell's world, are built around holes."

At the center of Russell's theory is a hollow point. Since there is no next-to in the modern standard of continuity, every point is surrounded by holes. That this notion would infect his account of perceived objects only shows his relentless consistency. What was ridiculous to Lovejoy seemed likable to Russell.

(In other words, for Russell, our world is not very different from his solar fantasy where intelligent beings have no solid bodies, but, more or less, their minds are scattered about by the winds, but still possess some consciousness of order).

6. Conclusion

Bertrand Russell championed the modern idea of continuity which he had received from Georg Cantor. In simplest terms, it states that between any two points, another can be inserted. This is not confined to the points within a line. It includes all possible directions surrounding a point – and even within the directions themselves, for they too are lines.

Russell tried to circumvent considerations of distance by going for topology, which concerned mainly with order. He tried to trivialize irrational numbers practically out of existence, reducing them to a lack or absence of boundary. . . And what is boundary for Russell and the technocrats who went along with him if it is not simply that which fits into their general idea of the world? The irrational number, they can neither grasp nor even touch. – but only approximate. Yet, his hero, Cantor, tried to prove that they were more numerous than rational numbers. Actually, Cantor's proof was incorrect, even on the premise of infinitely long digits. However, there is no doubt of the magnitude of them is great within a defined unit. length. That means that the measurements numbers made by scientists all too often result in blurs.

I have shown the insuperable difficulties caused with respect to the determination of what a point would have to be, the pathetic imitations of necessity and impossibility, the hall of mirrors attempt at a an empirical idea of what constitutes an object.

Russell was adamant: "Kant's antinomies, and the supposed difficulties of infinity and continuity, were finally disposed of by Georg Cantor. But at what a price! Coherence was certainly lost. The supposed solution to the problem of infinity offered by Cantor and accepted by Russell asserted that a part of an infinite whole can be as great as the whole itself -- another contradiction.

There is a notion of continuity which does not succumb to the pits and paradoxes laden within the Cantorian idea. It also holds that one of Kant's antinomies is not necessary; but it affirms, not denies space. Those who don't want to carry the modernist suppositions in the manner of a gang member sporting his special nick or tattoo will find this notion in my book, The Nature of Infinitesimals. As is stated in that work: "The modernist idea of continuity is backwards. The proof of continuity should be that nothing could be inserted between two neighbors, not that it always can be. The continuity of the number system is not the same as always keeping busy."

References

[1] Marquardt, Peter, the Natural Philosophy Alliance website, http://www.worldnpa.org/site/our-minimal-concensus/

[2] 1 Loc. cit.

1 Loc. cit.

1 Ibid., p. 72.

1 Ibid.. p. 73.

1 Russell, The Analysis of Matter, pp. 58, 279.

1 Ibid., p. 299.

1 Ibid., p. 295.
1 Ibid., p. 298.
1 Loc. cit.
1 Loc. cit.
1 Ibid., p. 312.
1 Ibid., p. 304.
1 Russell, "Reply to Criticisms," op. cit., p. 697.
1 Brand Blanshard, Reason And Analysis, (La Salle, 1964), pp. 136-7.
1 Lillian G. Lieber and Hugh Gray Lieber, Mits Wits and Logic, (New York: W. W. Norton & Co., 1960), p. 215.
1 Blanshard, op. cit., p, 159.
1 Ibid., p. 166.
1 Ibid., pp. 166-7.
1 Ibid. p. 167-8.
1 Bertrand Russell, "My Intellectual Development," The Philosophy of Bertrand Russell, p. 8.
1 Russell, The Analysis of Matter, p. 279.
1 Bertrand Russell, The Analysis of Mind, p. 98, quoted in Arthur O. Lovejoy's The Revolt From Dualism, (La Salle, Ill: The Open Court Publishing Company, 1930, 1960), pp. 239-40.
1 Loc. cit.
1 Bertrand Russell, "Perception and Physics," in Mind, N.S. (1922), p. 483,; quoted in Lovejoy, op. cit., pp. 243-5.
1 Bertrand Russell, The Analysis of Mind, p. 108; quoted in Lovejoy, op. cit., p. 241.
1 Lovejoy, op. cit., pp. 243-4.
1 Ibid., pp. 244-5.
1 Erickson, op. cit., pp. 53-7.
1 Russell, The Analysis of Matter, p. 14.
1 Erickson, op. cit., p. 44.

Accelerating Clocks Run Faster and Slower

Raymond HV Gallucci, PhD, PE;
8956 Amelung St., Frederick, Maryland, 21704;
gallucci@localnet.com, r_gallucci@verizon.net

Einstein's relativity contends that time, as measured by clocks, slows with increasing speed, becoming especially noticeable as the speed of light is approached. Discussions of this usually focus on constant speeds, albeit near the speed of light, and phenomena such as muon decay (near light speed), or even the Hafele-Keating experiment (at much slower speeds), are cited as 'proof.' Dissident scientists often contend that time remains invariant, although clocks may appear to run slower at increasing speeds. At least one such scientist contends that accelerated clocks can run both slower and faster, an interesting departure that I decided to examine via some examples. To the extent that my examples are correct, I too would agree with this conjecture, namely that, while time remains invariant, clocks can run faster and slower when accelerated (but not at constant velocity).

1. Introduction

While perusing Don E. Sprague's website on "Complex Relativity" (http://complexrelativity.com), I read the following discussion:[1]

> Clocks lose time but also gain time. *The Hafele and Keating experiment has atomic clocks going around the world showing less time in one direction but time gain in the other direction. We know that Einstein predicts that time slows with movement and eventually time is varied to a singularity where time end which is an impossibility. Since Einstein predicts that time slows, the Hafele and Keating experiment refutes Einstein. The clocks in the Hafele and Keating experiment show both a time loss and a time gain. According to Einstein, they just have time loss. Thus, the time gain portion goes against Einstein. However; the clock gain and loss is accurately predicted using CM* [Classical Mechanics] *and ChR* [Classical hierarchy Relativity] *with relative c. That is because ChR specifies that acceleration of a clock will result in a clock change in reading or clock error. Any examination of the Hafele-Keating experiment must consider the total acceleration of the clocks as they relate to the known universe.*
>
> *Consider an atomic clock experiment with the clock moved up a foot and down a foot resulting in a clock reading variation or error. This acceleration of the clock caused a loss of synchronization in the clock as predicted in ChR. The combination of the Hafele and Keating and the atomic clock one foot elevation experiments are confirmation that Maxwell/Einstein constant c relativity is wrong. It is proof of that ChR with relative c is correct.*
>
> *The combination of the Hafele and Keating experiment and the atomic clock 1 foot acceleration could loosely be considered to be the ChR equivalent of the Eddington observation about Einstein's relativity where he interpreted a gravitational lens bending light as confirmation that the time changed. In the case of the accelerating clocks, there isn't any way to interpret the clock gain as conformation of Einstein that predicts just time loss. There can only be clock error with accelerated clocks as specified in ChR.*
>
> *It isn't a matter of if Einstein is wrong while CM and ChR with constant space and constant progression of time and relative speed of light is correct in a hierarchy of frame relativity. It is just a question of when and how the physics world will acknowledge the truth I have shown.*

Others have disputed the contention that the Hafele-Keating results support Einstein's relativity (e.g., Spencer and Shama, "Analysis of the Hafele-Keating Experiment," Third Natural Philosophy Alliance Conference, Flagstaff, Arizona, June 1996; Kelly, "Hafele & Keating Tests: Did They Prove Anything?" [http://www.anti-relativity.com/hafelekeating debunk.htm]). Never being one to accept Einstein's conjecture that time slows due to movement at constant velocity, I nevertheless never considered the possibility of clocks (not time) showing variation under accelerated movement. The above discussion prompted me to consider this possibility by postulating three examples of acceleration: (1) change in speed, but not direction; (2) change in direction but not speed; and (3) change in both speed and direction. As my 'clock,' I postulate a gun shooting a projectile into a target, with the time between ejection from the gun and striking of the target becoming the unit of time measurement.

2. Case 1. Acceleration due to Change in Speed but not Direction

In Figure 1, a boxcar of length two (arbitrary units) has a pair of guns (grey) mounted to fire in opposite directions at its midpoint (shown here as 'upper' and 'lower'). At time 0, when the boxcar is stationary, both guns fire projectiles at equal speeds of $u_0 = 1$/sec (s). At an infinitesimal time later (0+), the boxcar, and therefore the two fixed guns, is accelerated to the right at $a_{0+} = 1/s^2$ (white arrows). Since both projectiles have already left their guns, neither 'feels' this acceleration, so each continues on its path at the original, constant speed. After 1 s, the boxcar has traveled $x = (1/s^2)(1\ s)^2/2 = 0.5$ to the right, now also the positions of the two guns (now with speeds of $v_1 = [1/s^2][1\ s] = 1/s$ to the right). Relative to their starting points in the boxcar, the projectiles have now reached the following positions: lower at +0.5, upper at −1.5 (having passed through the left wall of the car).

When stationary, an observer measures the 'standard' unit of time on the boxcar as that for a projectile to reach a wall, the same for each gun-projectile system. However, now the accelerated observer, assuming equal-speed projectiles, would conclude a clock calibrated to the upper gun runs faster than one calibrated to the lower gun because its projectile reaches a wall sooner – and that the

[1] The author acknowledges correspondence with Don E. Sprague regarding his theory and my development of the three examples.

upper clock runs faster than 'standard' time while the lower one runs slower. Direction matters.

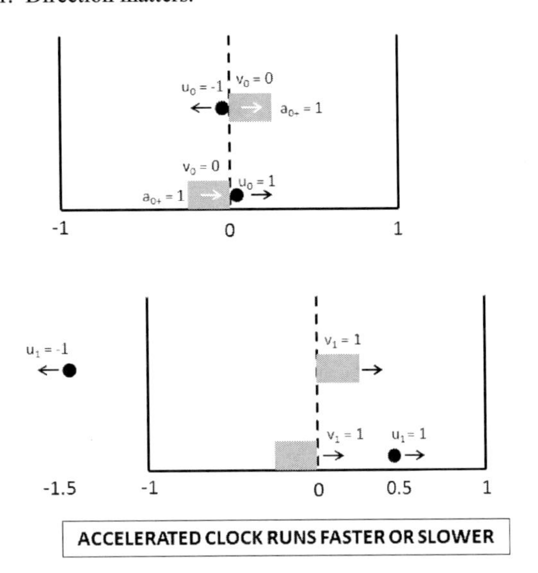

FIGURE 1. Case 1 – Boxcar Accelerating in Speed Only, not Direction

3. Case 2. Acceleration due to Change in Direction but not Speed

For the next two cases, it is convenient to examine circular motion, as that inherently involves directional acceleration and, if rotational speed is changed, acceleration in speed as well. First, we consider the case of acceleration due only to directional change, as shown in Figures 2 and 3. In Figure 2, a carousel (torus) rotates at a constant speed of 2π radians/s, such that the tangential speeds v_t of the inner and outer rims are 2/s and 6/s, respectively, given the radii shown (in arbitrary length units). A grey gun fixed to the inner rim, with its end rotating at v_t = 2/s, shoots a projectile from Point 0 at radial speed $v_r = (100/\pi)$/s such that it travels at speed $v = ([2/s]^2 + [\{100/\pi\}/s]^2)^{0.5} = 31.89$/s at angle $\alpha = \arctan(2/[100/\pi]) = 0.06275$ radian (3.595°). It follows Path 0-B to hit the outer rim at Point B after traveling a length of $\{2\cos(\pi-\alpha) + ([2\cos(\pi-\alpha)]^2 + 32)^{0.5}\}/2\pi = 0.6370$, using the law of cosines. The elapsed time is (0.6370)/(31.89/s) = 0.01997 s. Point A, on the outer rim, immediately above the gun, rotates to Point A' = (0.01997 s)(2π radians/s) = 0.1255 radian (7.191°) from the original Point A. Point B corresponds to rotation by arccos $\{(\pi^2/6)\ (10/\pi^2 - 0.6370^2)\}$ = 0.04185 radian (2.398°).

Define a new time unit, the 'zek' (z), as the time for the projectile to hit the outer rim. When stationary, one z = $(3/\pi - 1/\pi)/([100/\pi]/s)$ = 0.02 s. When rotating as shown, one z = 0.01997 s, i.e., 'time' appears to have sped up by (0.02 – 0.01997)/0.02 = 0.001313 (~0.13%). But really time has not varied; only the directional acceleration has caused an apparent speeding up by ~0.13%. If we use the projectile hitting the outer rim as a clock and standardize it when the carousel is stationary (one z), we conclude that, when accelerated, the clock runs faster (1 + 0.001313 = 1.001313 z by the standard clock).

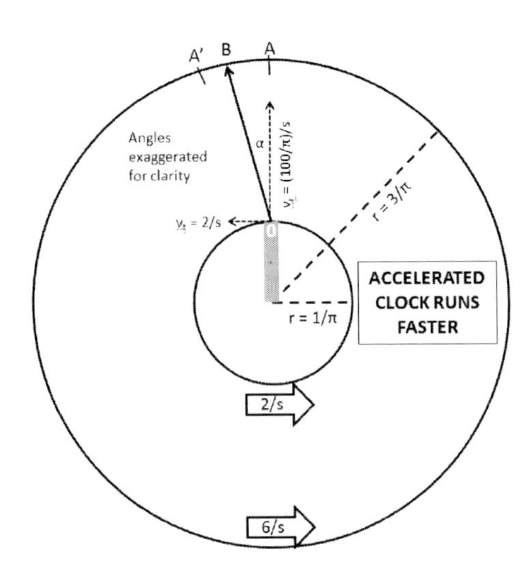

FIGURE 2. Case 2 – Carousel Rotating at Constant Speed with Gun Mounted on Inner Rim – Directional Acceleration Only

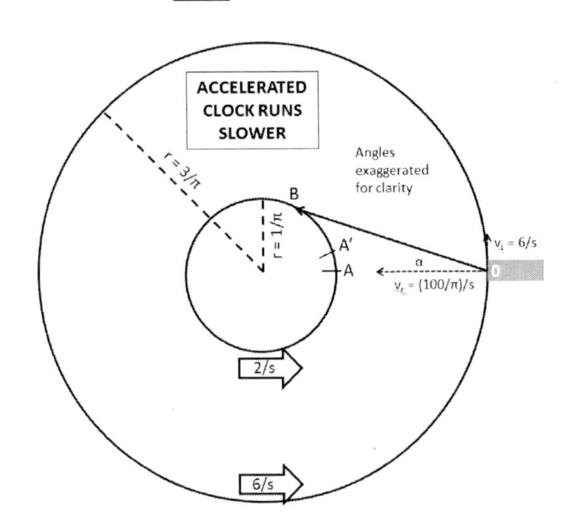

FIGURE 3. Case 2 – Carousel Rotating at Constant Speed with Gun Mounted on Outer Rim – Directional Acceleration Only

Figure 3 is the same as Figure 2, but now with the gun mounted on the outer rim. With its end rotating at v_t = 6/s, it shoots a projectile from Point 0 at radial speed $v_r = (100/\pi)$/s such that it travels at speed $v = ([6/s]^2 + [\{100/\pi\}/s]^2)^{0.5} = 31.93$/s at angle $\alpha = \arctan(6/[100/\pi]) = 0.1863$ radian (10.67°). It follows Path 0-B to hit the inner rim at Point B after traveling a length of $\{6 \cos\alpha - ([6 \cos\alpha]^2 - 32)^{0.5}\}/2\pi = 0.6738$, again using the law of cosines. The elapsed time is (0.6738)/(31.93/s) = 0.02111 s. Point A, on the inner rim, immediately below the gun, rotates to Point A' = (0.02111 s)(2π radians/s) = 0.1326 radian (7.598°) from original Point A. Point B corresponds to rotation by arccos $\{(\pi^2/6)\ ([10/\pi^2 - 0.6738^2)\}$ = 0.4029 radian (23.08°).

Now define the zek (z) as the time for the projectile to hit the inner rim. When stationary, one z again = 0.02 s. When rotating as shown, one z = 0.02111 s, i.e., 'time' appears to have slowed by (0.02111 – 0.02)/0.2 = 0.05523 (~5.5%), an opposite effect. But really time has not varied; only the directional acceleration has caused an apparent slowing by ~5.5%. If we again use the projectile hitting the inner rim as a clock and standardize it when the carousel is stationary (one z), we conclude that, when accelerated, the clock runs slower (1 - 0.05523 = 0.94477 z by the standard clock). As with Case 1, direction matters.

4. Case 3. Acceleration due to Change in Both Speed and Direction

For the final two cases, we continue with our rotating carousel, but now with the addition of acceleration in rotational speed. In Figure 4, the carousel rotates as before, with the grey gun mounted on the inner rim shooting a projectile as before. However, now at an infinitesimal time later (0+), the carousel is accelerated at 2π radians/s^2, such that the tangential accelerations a_t of the inner and outer rims are 2/s^2 and 6/s^2, respectively (grey arrows). The projectile does NOT experience this acceleration and, as before (Figure 2), reaches the outer rim in 0.01997 s. Because the carousel now speeds up, it will rotate by [4π radians/s + (2π radians/s^2)(0.01997 s)](0.01997 s)/2 = 0.1268 radian (7.262°), such that the projectile strikes the outer rim at Point B', with a perceived trajectory 0-B' now of length $[(10 - 6\cos[0.1268])/\pi^2]^{0.5} = 0.6404$.

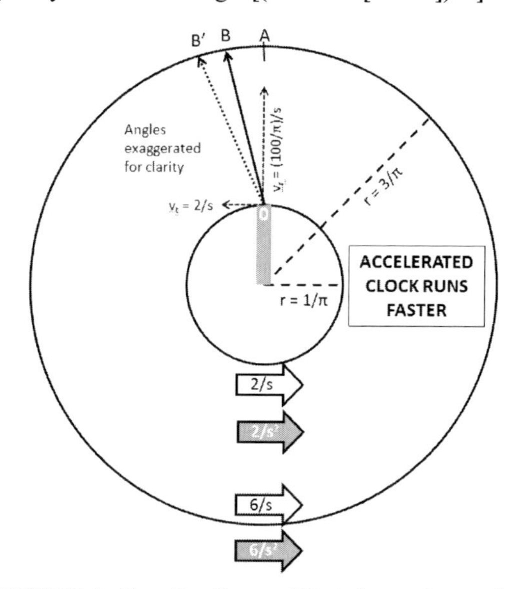

FIGURE 4. Case 3 – Carousel Rotating at Increasing Speed with Gun Mounted on Inner Rim – Both Speed and Directional Acceleration

When the carousel was not speeding up, the trajectory 0-B length was 0.6370 and required 0.01997 s (1.001313 z) to reach the outer rim. Now the length (trajectory 0-B') is longer (0.6404) and requires 0.6404/([100/π]/s) = 0.02012 s, or ([1.001313 z][0.02012 s]/[0.01997 s]) = 1.008644 z, to reach the outer rim. That is, more time has elapsed, which means the additionally accelerated clock (speed plus direction) now runs faster by (1.0086443 – 1.001313)/(1.001313) = 0.007321 (~0.73%).

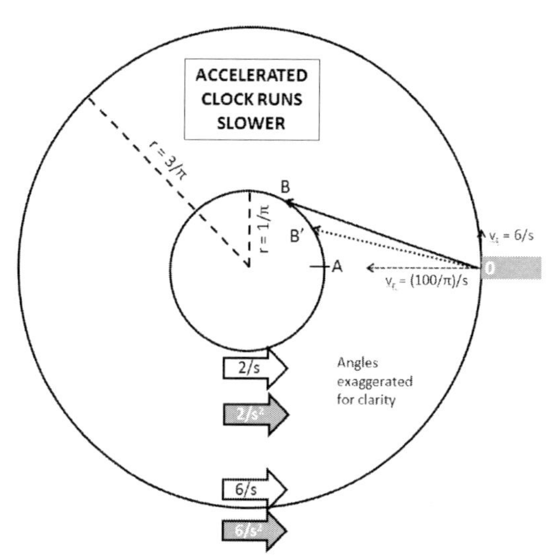

FIGURE 5. Case 3 – Carousel Rotating at Increasing Speed with Gun Mounted on Outer Rim – Both Speed and Directional Acceleration

Figure 5 is the same as Figure 4, but now with the grey gun mounted on the outer rim with its end rotating at v_t = 6/s. Again, at an infinitesimal time later (0+), the carousel is accelerated at 2π radians/s^2, such that the tangential accelerations a_t of the inner and outer rims are 2/s^2 and 6/s^2, respectively (grey arrows). The projectile does NOT experience this acceleration and, as in Figure 3, again reaches the inner rim in 0.02111 s. Because the carousel now speeds up, it will rotate by [4π radians/s + (2π radians/s2)(0.02111 s)](0.02111 s)/2 = 0.1340 radian (7.677°), such that the projectile strikes the inner rim at Point B', with a perceived trajectory 0-B' now of length $[(10 - 6\cos[0.1340])/\pi^2]^{0.5} = 0.6409$.

When the carousel was not speeding up, the trajectory 0-B length was 0.6738 and required 0.02111 s (0.94477 z) to reach the inner rim (remember the zek has different durations based on direction). Now the length (trajectory 0-B') is shorter (0.6409) and requires 0.6409/([100/π]/s) = 0.02013 s, or ([0.94477 z][0.02013 s]/[0.02111 s]) = 0.90132 z, to reach the inner rim. That is, less time has elapsed, which means the additionally accelerated clock (speed plus direction) now runs slower by (0.94477 – 0.90132)/(0.94477) = 0.04599 (~4.6%). Again, as with Cases 1 and 2, direction matters.

5. Conclusion

Can accelerating clocks run both faster and slower? Sprague believes so and provides his arguments on his website. I endeavored to examine this possibility using three cases considering both speed and directional changes as part of acceleration. As a result, I come to the same conclusion. This does not imply any belief in the variation of time itself, whether under constant or accelerating velocities, but merely a physical effect on an accelerating 'clock.' It also does not imply any belief that a clock moving at a constant velocity, even near the speed of light, will show any variation. The key is acceleration. And direction matters.

Michelson-Morley Interferometer Experiment of 1887: "Null" Result

Raymond HV Gallucci, PhD, PE;
8956 Amelung St., Frederick, Maryland, 21704
gallucci@localnet.com, r_gallucci@verizon.net

The Michelson-Morley Interferometer Experiment of 1887 is often cited as one of the cornerstones (and perhaps THE cornerstone) upon which Einstein built his theory of special relativity. Allegedly, it "proved" there was no aether. Once Einstein postulated that the speed of light was invariant, the only explanation that became accepted was that time slowed and length contracted due to relative motion according to the Lorentz Transformation formulae, adopted by Einstein as tenets of his special relativity. Despite subsequent experiments contradicting the alleged "null result," reanalysis of the results indicating positive ("non-null") results, and even maintaining the validity of the null result but explaining it via classical physics, the M&M Interferometer Experiment remains a special relativity foundation. However, if the limitation of the invariance of the speed of light is removed, the "null result" can be easily explained without resort to special relativity and its postulates of time dilation and length contraction. Yet this is seldom done.

1. Introduction

As described in "Michelson-Morley (M&M) experiment" (*http://en.wikipedia.org/wiki/Michelson%E2%80%93Morley_experiment*):

> *The Michelson–Morley experiment was published in 1887 by Albert A. Michelson and Edward W. Morley and performed at what is now Case Western Reserve University in Cleveland, Ohio. It compared the speed of light in perpendicular directions, in an attempt to detect the relative motion of matter through the stationary luminiferous aether ("aether wind"). The negative results are generally considered to be the first strong evidence against the then-prevalent aether theory, and initiated a line of research that eventually led to special relativity, in which the stationary aether concept has no role. The experiment has been referred to as "the moving-off point for the theoretical aspects of the Second Scientific Revolution" ... Together with the Ives–Stilwell and Kennedy–Thorndike experiments, the Michelson–Morley experiment forms one of the fundamental tests of special relativity theory.*

Subsequent experiments have called into question the need for special relativity to explain the alleged "null result" (e.g., *http://www.anti-relativity.com/daytonmiller.htm;http://www.relativityoflight.com/Chapter9.html;http://www.neoclassicalrelativity.org/;http://www.conspiracyoflight.com/M&M.html;http://www.orgonelab.org/miller.htm*).[1] Even reanalysis of the M&M results has suggested that there was a fringe shift, contrary to the alleged "null" result (*http://relativitychallenge.com/papers/Bryant.CICS.MMX.Analysis.06302006.pdf;* R. Cahill, "The Michelson and Morley 1887 Experiment and the Discovery of Absolute Motion," *Progress in Physics,* October 2005, Volume 3, pp. 25-29).

2. Analysis

Relativistic length contraction (time dilation) is usually cited as the explanation for the "null result of the famous 1887 M&M Interferometer Experiment that reputedly prompted Einstein's Special Relativity. However, if we allow that light can travel at velocities other than c, a much simpler explanation is available.

As shown in Figure 1, the M&M Interferometer Experiment effectively sent two perpendicular light rays from a source (solid mirror) to two target mirrors (shaded and hollow), each a distance L away from the source at time step 0 while the apparatus was translating along one of the ray lines at speed v (presumably that of the Earth tangentially relative to the Sun). Since the source is moving at v, the speeds (black arrows) of the light rays (dashed, dotted and mixed) in the vertical and horizontal directions are vector sums of c and v, i.e., $(c^2+v^2)^{0.5}$ vertically and (c+v) horizontally. The distances (scalars [no arrows]) traveled over time step 1 (at which time "t" the perpendicular ray strikes the shaded mirror and the horizontal ray strikes the hollow mirror) are $(L^2+[vt]^2)^{0.5}$ and (L+vt), respectively.

By symmetry, from time step 1 to 2, the two rays (dashed-dotted and mixed) are reflected back to the source mirror over another time "t." The perpendicular ray covers the same distance at the same speed. However, the horizontal ray now covers a shorter distance (L+vt-2vt = L-vt) at a slower speed (c-v). Since the time "t" is equal in both directions for each time step, we can express it as follows (time = distance/speed):

$$\{(L^2+[vt]^2)^{0.5}\}/\{(c^2+v^2)^{0.5}\} = (L+vt)/(c+v) = (L-vt)/(c-v) = (L\pm vt)/(c\pm v).$$

Squaring both sides yields $(L^2+v^2t^2)/(c^2+v^2) = (L^2\pm 2vt+v^2t^2)/(c^2\pm 2cv+v^2)$, which, after "cross-multiplying" and dividing by 2v, simplifies to $L^2c + cv^2t^2 = Lc^2t + Lv^2t$.

[1] Note that the citing of these various websites does not necessarily imply the author's agreement with all material presented on the site. These are cited solely for the portions of their discussions related to the M&M Interferometer Experiment.

This can be more simply expressed as $v^2t(ct - L) = Lc(ct - L)$. Since there is no *a priori* reason for v^2t to equal Lc, the only way this equation can hold is if both sides are zero, i.e., L = ct. But this is precisely the situation governing the relationship for light propagation between the source mirror and each of its target mirrors relative to the three mirrors (and the apparatus as a whole, i.e., the "moving" system). That is, over either time interval "t," the source mirror (or, equivalently, each target mirror) sees the light ray(s) cover the distance L vertically or horizontally at speed c. Therefore, the time elapsed in either the "stationary" (relative to the Sun) or "moving" (relative to the apparatus) reference frame is the same ("t"). There is no time or length dilation, no relativistic effects – therefore, the (in?)famous "null result."

3. Conclusion

Does special relativity, via the Lorentz Transformations, explain the alleged "null result" from the M&M Interferometer Experiment of 1887? Yes. Is that theory and those transformations the only possible explanation? No. Other dissident physicists have offered various non-relativistic explanations of the results, at least one alleging a fringe shift occurred, contradicting the "null result." I too offer a simple classical explanation, based on relaxing the limitation of the invariance of the speed of light, allowing light to acquire the velocity of its source.

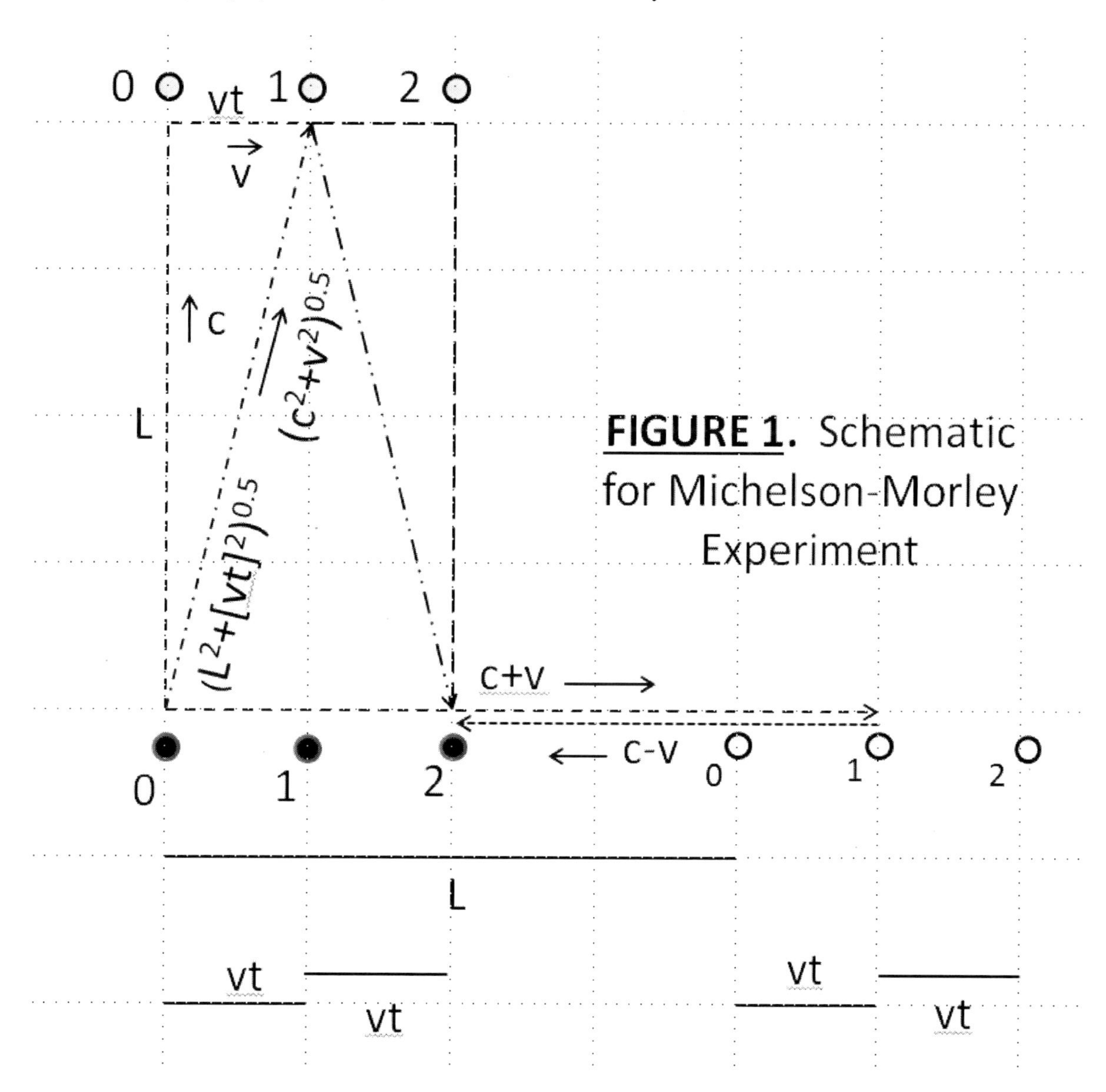

FIGURE 1. Schematic for Michelson-Morley Experiment

Questioning the Cosmological Doppler Red-Shift

Raymond HV Gallucci, PhD, PE;
8956 Amelung St., Frederick, Maryland, 21704
gallucci@localnet.com, r_gallucci@verizon.net

Linked with the concept of a cosmological Doppler red-shift is an expanding universe with rapidly receding stars, galaxies, etc. Assuming no form of matter, especially a reasonably macroscopic and tenuous one like a star, let alone an entire galaxy, could possibly travel at speeds approaching that of light and remain 'intact' (except, perhaps, something as dense as a neutron star), the only possible way for such an entity to exhibit recession speeds approaching that of light would be for space itself to be expanding. And whether one accepts the traditional or Galilean Doppler red-shift as the correct explanation, one is still left to conclude that 'something' is 'racing away.' I endeavor to cast doubt on the traditional explanation of a cosmological Doppler red-shift due to universe expansion. A Galilean Doppler red-shift may be an equally plausible explanation for those who adhere to the premise of stars, galaxies, etc., receding for whatever reason.

1. Introduction

Earth, effectively a stationary observer, receives red-shifted light from a star speeding away at 0.2c. The amount of red-shift, i.e., frequency reduction/wavelength increase, based on traditional Doppler Effect formulae, is c/(c + 0.2c) = 0.833 (frequency reduction) or (c + 0.2c)/c = 1.2 (wavelength increase). If viewed as a cosmological red-shift, the corresponding 'z' value is (1.2 – 1)/1 = 0.2 [or, (1 – 0.833)/0.833 = 0.2]. All these calculations assume light travels at the fixed speed c and any red-shift is due to a change in the waveform (i.e., shape of the wave, analogous to stretching a spring), not a change in light speed. Other types of waves (sound, water, etc.) propagate as pulses through a medium, where the matter (e.g., air or water molecules) interact with each other to produce and propagate the pulse. Except for some longitudinal (sound) or transverse (water) movement among the matter, the matter itself essentially 'remains in place.' These waves also have fixed propagation speeds, given fixed physical properties such as temperature, pressure, density, viscosity, etc., and form the basis for the Doppler Effect formulae.

Light has been assumed to behave similarly, i.e., traveling at a fixed speed, again depending upon the medium and its properties, although for the case of light in a vacuum, no medium is deemed necessary. In fact, the only effect of a medium (e.g., interstellar or intergalactic 'dust,' gas or plasma) is to retard the light speed relative to its maximum possible value of c in a vacuum which, by definition, is void and thereby cannot constitute a medium. Unless one returns to the 19th century concept of the aether (and many who question the validity of Einstein's relativity have done so), what is the basis for assuming that light, if a wave, behaves in the same way as waves whose transmission is dependent upon the interaction of matter in a medium that can transmit pulses?

One could argue that such a wave (e.g., sound or water) is nothing more than movement of the medium itself, albeit longitudinally (sound) or transversely (water). Without the medium, there can be no wave. Therefore, unless there is an aether, how can a light 'wave' be analogous to any other type of wave? And (at the risk of sounding somewhat 'McLuhanish'), would it not be the fact that "the medium itself is the wave" is what results in constant wave speeds for sound or water, given fixed physical properties? Therefore, unless light is the movement of a medium (aether?), why should it have a fixed speed?

Furthermore, given the previous, why should the traditional Doppler Effect apply to light, especially given that it is considered more than just a wave, i.e., the quantum wave-particle duality? And, if there is such a medium as an aether, why is it totally undetectable, other than theoretically its effect on light (or its movement being light itself)? It would appear to have no physical properties, consist of no form of matter, etc. And, if there is truly no aether, and light has no transmitting medium, why assume its wave behavior is analogous to waves propagating through a medium (e.g., fixed transmission speed, subject to traditional Doppler Effect)?

2. An Alternative

Consider another possibility, assuming that light retains some aspect(s) of traditional wave-like behavior. Rather than the effect of the star receding at 0.2c being on wavelength and frequency, what if it acts directly to reduce the light speed to 0.8c (Galilean transform), contrary to relativity, constant speed of light, etc.? Assume the waveform (shape) remains unaffected, i.e., the frequency and wavelength are the same (no stretching), but only the transmission speed changes (is reduced). Whereas a stationary star would emit light at speed c, such that over an increment of time 't,' 'n' wavelengths (cycles) of length 'w' (m/cyc) would be received at Earth (implying a frequency 'f' = n/t), the receding star would emit light at speed 0.8c, reducing the number of wavelengths received over t to 0.8n (with reduced frequency = 0.8n/t). Note that, calculationally, this is equivalent to the traditional Doppler effect on frequency, but now with the source (star) stationary and the observer (Earth) receding at 0.2c. Traditionally, the Doppler Effect in this case is (c - 0.2c)/c = 0.8 (frequency reduction), the same value, although quite conceptually different (see Figure 1, last page of this article).

Traditionally, the 'length' of the transmission is unaffected by the Doppler shift, in that it remains the same whether the star is stationary or receding: stationary length = w (m/cyc) x f (cyc/sec) x t (sec) = wft (m); receding length = 1.2w (m/cyc) x 0.833f (cyc/sec) x t (sec) = wft (m); both of these = ct. If the light speed is affected rather than the waveform, the receding length is shortened as follows: w (m/cyc) x 0.8f (cyc/sec) x t (sec) = 0.8wft (m), i.e., 0.8ct. The corresponding cosmological Doppler Effect z value is (1 – 0.8)/0.8 = 0.25, slightly higher ('redder') than that from the traditional approach. Unfortunately, since it is z that is used to calculate the star's recession speed, there would need to be another independent way of calculating that speed to determine which red-shift estimate (0.833f or 0.8f) is correct (if either).

For the case where the star is stationary, there is no Doppler red-shift and two full wavelengths (cycles) are shown in Figure 1 as traversing the length over time t, at frequency 2/t (cyc/sec). With the star receding at 0.2c, the traditional Doppler red-shift, with light still traveling at c, 'elongates' the wave such that only 2/1.2 = 1.67 wavelengths traverse the length over time t, at frequency (2 x 0.833)/t = 1.67 (cyc/sec). The corresponding cosmological red-shift is calculated as z = 0.2, from which one would infer the star's recession speed to be 0.2c. Lastly, with the star again receding at 0.2c, but now assuming the emitted light experiences this as a reduction in speed, rather than a change in waveform, a shorter length is traversed over time t, spanned by 2 x 0.8 = 1.6 wavelengths, at frequency = (2 x 0.8)/t = 1.6/t (cyc/sec). If this were assumed to correspond to the traditional Doppler red-shift, the inferred recession speed of the star would be based on z = (2 – 1.6)/1.6 = 0.25, i.e., 0.25c. However, if the Doppler effect is 'Galilean,' i.e., the speed of light transmission, not the waveform, is affected by the speed of the source, then the true speed of the star is 0.2c (recession).

3. Analogy with Refraction?

Consider the traditional view for light refraction when entering a denser (or less dense) medium, shown in Figure 2 (see last page) *(http://www.bing.com/images/search?q=refraction+of+light&qpvt=refraction+of+light&FORM=IGRE#view=detail&id=E1665849CC7E68512A240617ACC9D06A1507E84C&selectedIndex=54).* The speed slows (or increases), with the full effect of the speed change being carried by a corresponding change solely in the wavelength, i.e., the frequency remains the same. Therefore, there is no change in 'color' (using this term loosely to apply to non-visible light as well) since 'color' is determined solely by frequency. (It is a common misconception that 'color' can be equivalently characterized by wavelength or frequency, unless one is speaking solely of travel through the same medium, where the light speed is constant [for stationary source and observer] and, therefore, a change in one reciprocally changes the other. The fact that there is no 'color' change during refraction demonstrates that 'color' is really a function solely of frequency.)

Consider a similar situation where you are sitting by a swimming pool at midnight gazing vertically upward at the full Moon. Assume the Moon is made entirely of green cheese. Gazing upward through the atmosphere (and void of space between the Earth and Moon) for ten seconds, the Moon stays green. Now, take a deep breath and dive under the water to the bottom. Again look vertically upward at the full Moon for ten seconds. While the image may be blurred from your splash, the Moon remains green, i.e., no 'color' change. Although the speed of light decreased when entering the water (this time perpendicularly so there is no refraction), only its wavelength, not its frequency, dropped. Therefore, there was no 'color' change. In this case, transmission of light through different media, the speed change is carried by the wavelength, not the frequency.

The two cases are different, however. The Moon and Earth are essentially stationary with respect to one another (at least for the ten-second interval) and the light traverses two different media. The speed changes (considering the void-atmosphere as one medium and the water as the other), but this change is carried only by a change in wavelength, not frequency. There is no 'red-shift' (the Moon stays as green as ever). Now, with the star receding relative to Earth but the light not changing its medium (slight difference between Earth's atmosphere and void of space notwithstanding), a Galilean Doppler shift due to the decreased light speed is carried solely by a decrease in frequency. There is no wave stretching, i.e., the wavelength does not change. However, the length of the wave reaching the observer over the same time period is reduced, therefore corresponding to a frequency reduction and, therefore, a red-shift.

Absent the identification of a propagating medium for light, what is the basis for assuming the wave portion of its behavior is the same as that for waves propagating through tangible media? Why would the wave from a receding source necessarily stretch while that from an approaching source compress? If we do not assume light speed is constant in a given medium (or, in the case of interstellar or intergalactic space, no medium at all), why would the wave necessarily behave similarly? To factually determine which model (if either) is accurate, one needs to independently measure the star's speed, independently measure the red-shift, and then see which, if either, formula yields consistent results. Otherwise it is, as with most of the basis for relativity, more theoretical speculation than experimental foundation.

4. Conclusion

However, after having discussed all this and having proposed an alternate version of the cosmological Doppler red-shift, I must state that I do not believe in an expanding universe or even rapidly receding stars, galaxies, etc. Assuming no form of matter, especially a reasonably macroscopic and tenuous one like a star, let alone an entire galaxy, could possibly travel at speeds approaching that of light and remain 'intact' (except, perhaps, something as dense as a neutron star), the only possible way for such an entity to exhibit recession speeds approaching that of light would be for space itself to be expanding. And whether one accepts the traditional or Galilean Doppler red-shift as the correct explanation, one is still left to conclude that 'something' is 'racing away.' Therefore, I believe the correct explanation for the apparent expansion of the universe is one of the various 'tired light' theories (or one yet to be proposed), whereby light's interaction with interstellar and/or intergalactic media reduces its energy, resulting in a non-Doppler red-shift.

Various theories, such as gravitational 'de-energization,' Compton scattering, 'dust' absorption-re-emission, quantum electro-dynamical interactions, are among many that have been proposed. All reduce the light energy, which can be viewed traditionally as a decrease in frequency with (traditionally) or without ('Galileanly') a corresponding increase in wavelength. Either way, a red-shift (non-Doppler) occurs. My goal has been not to resolve which of these tired light theories is most plausible, but to cast doubt on the traditional explanation of a cosmological Doppler red-shift due to universe expansion. A Galilean Doppler red-shift may be an equally plausible explanation for those who adhere to the premise of stars, galaxies, etc., receding for whatever reason.

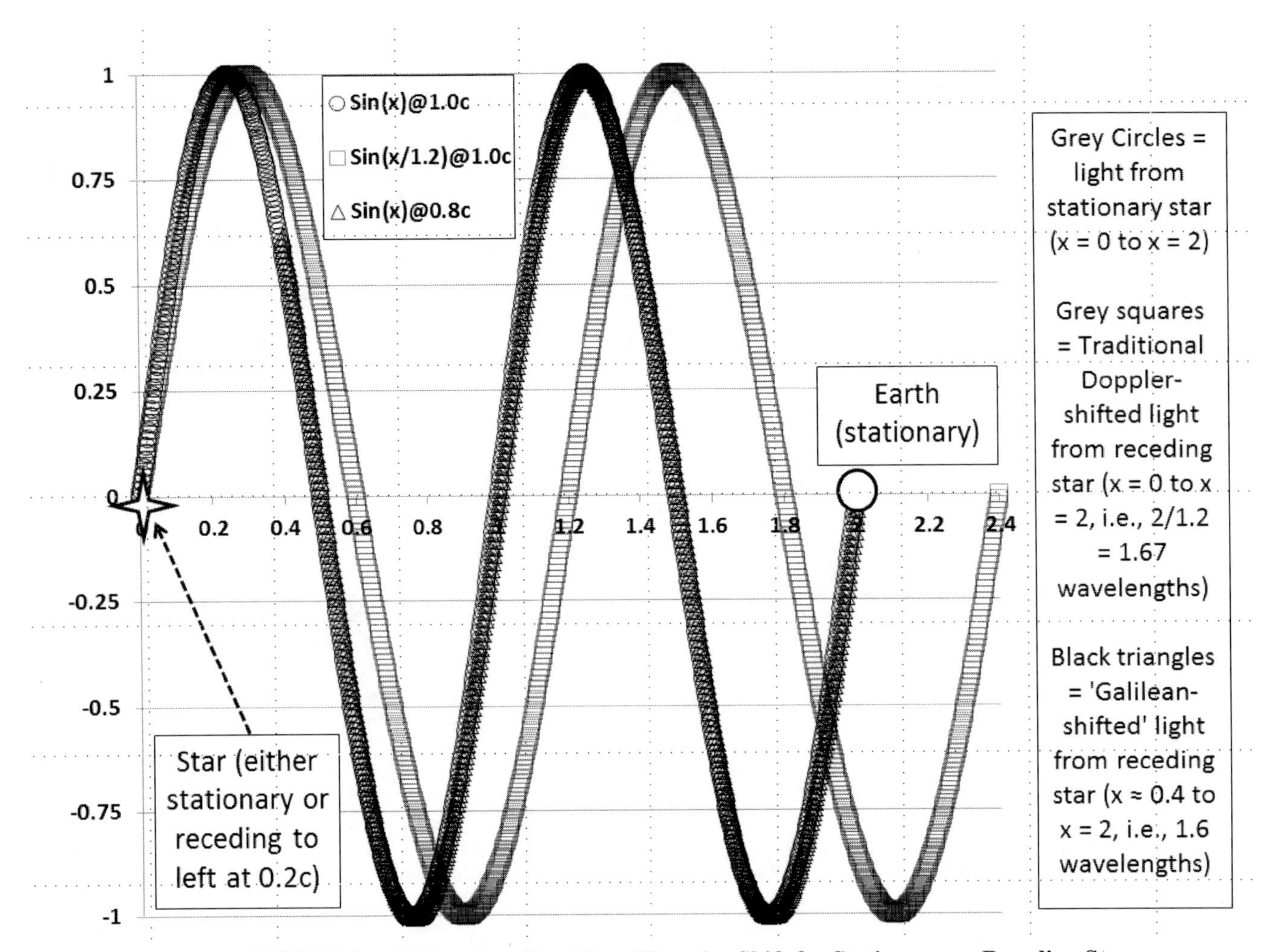

FIGURE 1. **'Galilean' vs. Traditional Doppler Shift for Stationary vs. Receding Star**

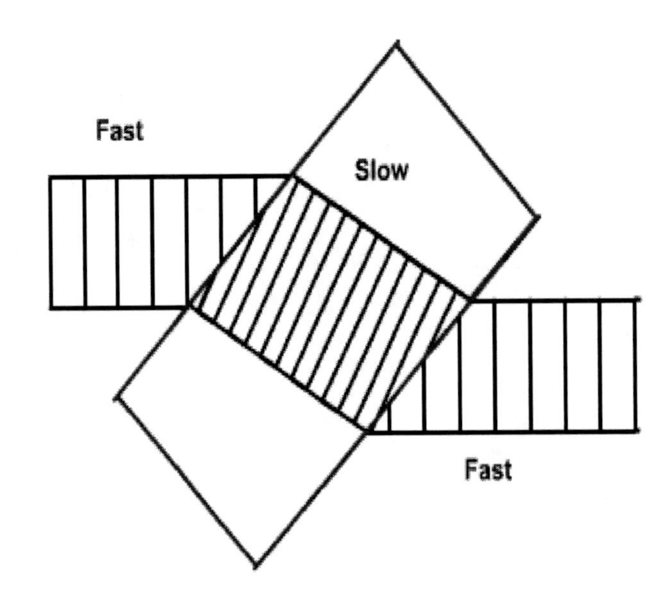

FIGURE 2. Light Refraction upon Passing Through a Denser Medium

Re-Examining Velikovsky

Raymond HV Gallucci, PhD, PE;
8956 Amelung St., Frederick, Maryland, 21704
gallucci@localnet.com, r_gallucci@verizon.net

Velikovsky's Worlds in Collision and other books inspired founders of the Electric Universe Theory with its implication of plasma-based electrical interactions between planets, as recorded in historical records such as the Bible. While his theories sparked great controversy by implying close encounters between some of the inner planets (Venus, Earth and Mars) within recorded human history, but were summarily dismissed as fiction by mainstream scientists and continue to be dismissed today. Nonetheless, Electric Universe theorists contend that there may be truth in Velikovsky's conclusions when these "interactions" are considered in light of a "plasma/electric universe." Having also read Velikovsky's works, and Electric Universe Theory, I endeavored to perform a simple set of calculations to ascertain if any of Velikovsky's interactions could have been possible.

1. Introduction

The "Electric Universe" Theory (EUT) owes part of its inspiration to the work of Immanuel Velikovsky (Worlds in Collision [1950]), at least for introducing the concept of catastrophism of an electrical nature potentially inducing what EUT proponents see as plasma-arced "scarring" on some planets and other objects within our solar system, such as Mars. Though discredited by mainstream physicists and astronomers since the publication of his ideas in 1950 (see Wal Thornhill, "The Impact of Pseudo-Science," March 17, 2000 [*http://www.holoscience.com/wp/the-impact-of-pseudo-science/*]):

> *In 1974, the AAAS [American Association for the Advancement of Science] held a session in San Francisco which was supposed to allow Velikovsky a forum to answer his critics. It was, as it transpired, a disgraceful ambush. Now, some quarter century later, the American Association for the Advancement of Science (AAAS) has discussed a similar topic but without Velikovsky's presence. The subject was "unpredictable events of extra-terrestrial origin and their impact on humanity". It was an occasion for the sensationalists to parade their predictions of doomsday by impact from a comet or asteroid. It also became another opportunity for academics to rewrite history and indulge in yet another miserable attack on Velikovsky. As reported in the WhyFiles: "...there are some neo-catastrophists, located mainly in Britain, who have an almost Velikovskian pseudo-scientific take on this matter and have argued that such impacts are more frequent ... Velikovsky, of course, is the guy who gave asteroid impacts such a bad name back in 1950."*

Velikovsky nonetheless ushered in an era where catastrophic events, rather than just immeasurable eons of uniformitarianism, became acknowledged as a potential contributor to the current state of our solar system (and maybe beyond). While EUT proponents acknowledge that "[i]t seems unlikely that Velikovsky's historical reconstruction of planetary catastrophes is correct," they also contend that

> *None of this denies Velikovsky priority in identifying the major destructive influence in the Earth's past as the near approaches of the planets Mars and Venus. His reconstruction of awesome celestial events in the dimly remembered past follow the laws of physics and the rules of evidence. His model is a good one when measured by its prediction score against that of conventional models. Conventional models are woefully deficient to pronounce upon impacts and their effects. To begin with, planetologists have admitted they are unable to experimentally reproduce the features of so-called impact craters. So, what are the craters? If they are not a result of impacts, what possible use are they in predicting future impacts? Is the science of impacts a pseudo-science?*

I agree that the EUT proponents' contentions regarding electrically plasma-arced scarring of certain planetary features such as those on Mars seem plausible. However, it takes quite a leap of faith to believe that, within human memory, a planetoid (or comet) the size of Venus careened through the solar system over a distance of at least ~ 4.5 Astronomical Units (~ 400 million miles, the distance of closest approach between the orbits of Venus and Jupiter [see Table 1]), encountering both the planets Mars and Earth before settling into its current orbit as the planet Venus. Certainly the perturbations induced by such an "astronomical" event would still have left all three planets "ringing" today. This is not to say that perhaps Mars and Earth (and perhaps even Venus) did not experience "close encounters" with some "rogue" planetoid or comet within human history that inspired the human memory cited by Velikovsky and caused the evidential scarring (e.g., see Allan and Delair, When the Earth Nearly Died [1997]). But to ascribe this to what is now the planet Venus seems too much to believe.

2. Analysis

Rather than dismiss this outright, I decided to perform some very simple, hopefully conservative, physics calculations to determine if such encounters could have happened within human memory. (Velikovsky proposed that Venus was ejected from Jupiter and had a close encounter with Earth roughly 3500 years ago [see "Immanuel Velikovsky" (*http://en.wikipedia.org/wiki/ Immanuel_Velikovsky*)]). While Velikovsky proposed multiple encounters between Earth and Venus, I will assume just one for my analysis. Table 1 lists the planetary properties that will be relevant for my calculations (retaining Pluto as a "planet").

If Venus were ejected from Jupiter as Velikovsky contends, it would have had to achieve at least the Jupiter Escape Speed of ~ 60 km/s (see "Escape Velocity" [*http://en.wikipedia.org/wiki/Escape_velocity#List_of_escape_velocities*]). An object escaping Jupiter will have a speed, depending upon its direction of escape, relative to the Sun of 60±13 km/s, since the latter is Jupiter's orbital speed. If Venus were expelled from Jupiter (which seems unlikely given the factor of four difference in their densities [see Table 1], but perhaps it was a very large moon that somehow was ejected), its slowest initial speed relative to the Sun would have been 47 km/s.

Ignoring any increase in speed as Venus accelerated inward toward the Sun, upon encountering Mars (presumably close enough to induced electrically plasma-arced scarring but not to have physically disrupted planetary integrity via tidal forces, i.e., no closer than the Roche limit of ~ 33,500 km, or about 2.5 times Venus' diameter), conservation of momentum would have required that

$$M_vS_{v1} - M_mS_{m1} = M_vS_{v2} - M_mS_{m2}$$

where M = mass and S = speed; for the subscripts, v = Venus, m = Mars, 1 = pre-encounter, 2 = post-encounter.

Note that I assume no change in mass for either planet during the encounter, only changes in their speeds. Further, I assume they pass each other going in parallel but opposite directions, to minimize the final speeds after the encounter (a slowing process). With the M values from Table 1, S_{v1} = 47 km/s (assuming the minimum ejection speed) and S_{m2} = 24 km/s (Mars' current orbital speed), the following relation evolves:

$$S_{v2} = 47 + (0.641/4.83)(24 - S_{m1})$$

Assuming Venus next encountered Earth (again, presumably no closer than the Roche limit of ~ 32,800 km, or ~ 2.5 times Earth's diameter), the same conservation of momentum would have required that (now with subscripts e = Earth and 3 = Venus' post-encounter with Earth)

$$M_vS_{v2} - M_eS_{e1} = M_vS_{v3} - M_eS_{e2}$$

again assuming no change in mass, no acceleration of Venus due to the Sun and an anti-parallel encounter. With the M values from above, S_{v3} = 35 km/s (Venus' current orbital speed) and S_{e2} = 30 km/s (Earth's current orbital speed), the following relation evolves:

$$S_{e1} = 30 + (5.2/5.52)[12 + (0.641/4.83)(24 - S_{m1})]$$

Unfortunately, we do not know S_{m1}, the initial speed of Mars, i.e., prior to its encounter with Venus. However, if we assume the original ("pre-Venus-encounter") speeds of Earth and Mars were in the same proportion as their current, we can derive

$$S_{m1} = (24/30)S_{e1}$$

Substituting this in the previous equation yields S_{e1} = 40 km/s, or ~ 33% faster than today. (The corresponding "pre-Venus-encounter" speed for Mars would have been 32 km/s, or also ~ 33% faster than today.) Thus, if Venus encountered Earth within human memory, there would be evidence of a pre-encounter "year" that was ~ 33% shorter than current, i.e., ~ 245 days.

Even if all these approximations and simplifications yield a result that is off by a factor of 10, a pre-encounter Earth "year" only ~ 3.3% shorter (353 days) than current would quickly accumulate into one of our current years in only ~ 30 years. Presumably even our ancestors would have noticed such a difference, as a "century" would have been over three "years" shorter than it is now. Coincidentally, a "year" shorter by ~ 3.3% (~ 12 days) has precedence regarding human reaction. In 1750, the Parliament of Great Britain switched from the Julian to the Gregorian calendar (other countries had previously converted as early as 1583), which had over time amounted to an 11-day difference. Thinking their lives had been shortened, the uneducated populace rioted for the "return of our 11 days." (By the time Russia switched in 1918, the difference had increased to 13 days.) (see "Calendar [New Style] Act 1750" [*http://en.wikipedia.org/wiki/Calendar_(New_Style)_Act_1750*]; and "Gregorian Calendar" [*http://en.wikipedia.org/wiki/ Gregorian_calendar*])

3. Conclusion

Based on my admittedly quite crude calculations, it still appears too far-fetched to believe the planet Venus arose out of Jupiter and careened through the inner solar system anytime within human history (if ever at all). However, this does not invalidate the EUT contention that planetary scarring due to electrical plasma-arcing may be responsible for the bizarre surface features seen on Mars and perhaps other celestial objects. But just what particular planetoids or comets might have been responsible for this remains unknown.

The Essence of the Universal Spacetime Theory

Vladimir B. Ginzburg
612 Driftwood Drive, Pittsburgh, PA 15238, USA
e-mail: helicola@aol.com

According to the UST. The toroidal spacetime entity called *toryx* is responsible for the creation of all kinds of matter in the Universe. What we perceive as matter, field, gravity, mass, charge and energy, etc. are merely various metamorphoses of polarized spacetime. Capable of existing in four topologically-polarized states, the polarized toryces are unified to form four kinds of elementary matter particles called *trons*, the building blocks of nucleons and atoms. Created by an excited toryx is the helical spacetime entity called the *helyx* responsible for the creation of all kinds of radiation particles in the Universe. The spacetime properties of both the toryx and the helyx are based on a limited number of assumptions, while their physical properties are readily expressed as the functions of their spacetime properties. The UST solves a century-long problem of the unification of physics of the micro- and macro-worlds. It presents the dark matter and dark energy as merely the metamorphoses of spacetime. Since the UST is 3-dimensional (plus time), its predictions can be verified by the experiments utilizing the 3-dimensional technology.

Introduction

The Universal Spacetime Theory (UST) is a summary of ideas that the author have been developing during since 1992 [1-20]. It attempts to answer four most intriguing and puzzling questions related to the science and philosophy:

1. How was the Universe created out of nothing?
2. Is there a single prime entity responsible for the creation of all kinds of matter in the Universe?
3. Is there a single prime entity responsible for the creation of all kinds of radiation in the Universe?
4. What are the common primordial laws governing the formation of ordinary matter, dark matter and dark energy?

The UST shows that it is possible to find answers to these questions by recognizing that what we perceive as matter, field, gravity, mass, charge and energy, etc. are merely various metamorphoses of polarized spacetime. Depending on a degree of its polarization, the spacetime may reveal itself either as "nothing" without any distinct properties known to us or as "something" having various detectable physical properties. The spacetime-based approach allowed the UST to discover a single prime entity responsible for the creation of all kinds of matter in the Universe. This entity is in the form of a toroidal spacetime called the *toryx*. Capable of existing in four topologically-polarized states, the polarized toryces are unified to form four kinds of elementary matter particles called *trons*, the building blocks of nucleons and atoms.

The spacetime-based approach also allowed the UST to discover a single prime entity responsible for the creation of all kinds of radiation in the Universe. This entity is in the form of a helical spacetime called the *helyx*. Emitted by the excited trons, the unified topologically-polarized helyces form four kinds of elementary radiation particles called *tons*. The tons are responsible for the communication between trons and some of them propagate with superluminal velocities. Finally, the spacetime-based approach allowed the UST to discover the primordial laws governing the formation of ordinary matter, dark matter and dark energy.

The idea of the spacetime playing an important role in the universe is not new. In 1915, the German-Swiss physicist Albert Einstein developed his general theory of relativity, in which he explained gravity as a geometric property of spacetime. However, while applicable to the macro-world of stars, galaxies, etc., this theory was found to be incompatible with quantum mechanics of the micro-world introduced in 1905 by the German physicist Max Planck. During the past 100 years many physicists have tried to unify these two theories by attempting to expand Einstein's spacetime into the micro-world, but with no success.

The UST solves this problem by using the same fundamental equations to describe the spacetime properties of both the micro- and macro-toryces. The functional roles of the micro- and macro-toryces, however, are different. The polarized micro-toryces create elementary particles forming the constituents of celestial bodies, while the polarized macro-toryces form the macro-spacetimes intimately associated with each celestial body and responsible for the interactions between the celestial bodies.

1. Basic Concept of the Toryx

As shown in Figures 1 and 2, toryx is the spiral spacetime entity made up of three parts:

- Circular *leading string* with the radius r_1 propagating along its circular path with the velocity V_1.
- Toroidal *trailing string* with the radius r_2 propagating with velocity of light c along its toroidal spiral path synchronously with the leading string.
- *Spherical membrane* with the radius r encompassing the trailing string.

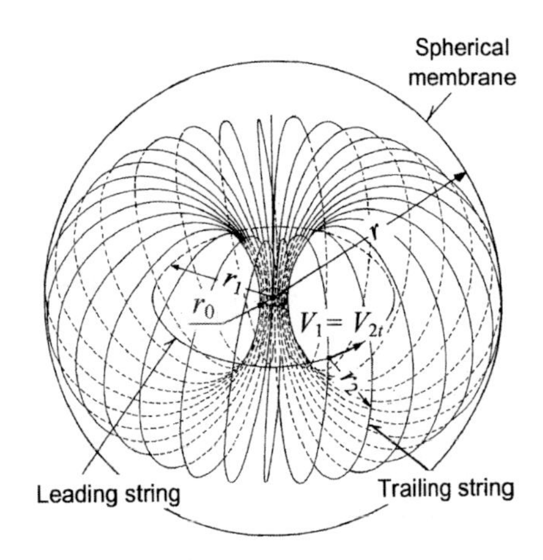

Figure 1. Isometric view of a toryx.

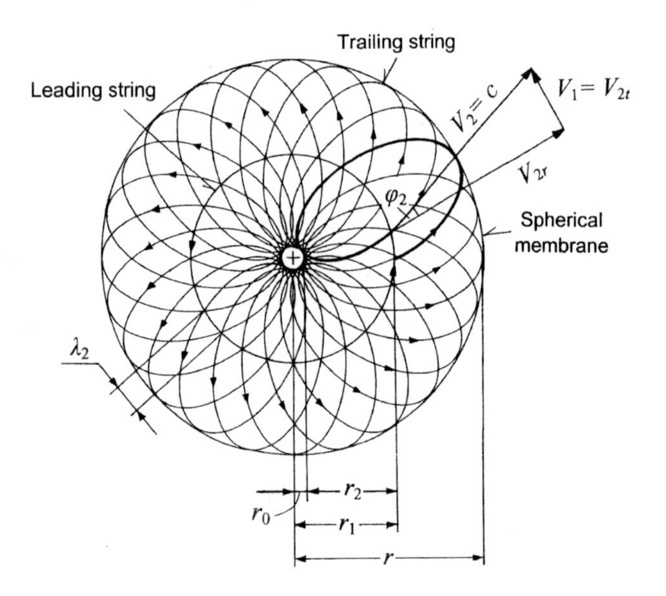

Figure 2. Top view of a toryx.

The toryx spacetime properties are based on three fundamental equations called the *toryx spacetime postulates*. These postulates limit the toryx degrees of freedom, allowing one to establish the relationships between all its spacetime parameters.

1. The length of one winding of the trailing string L_2 is equal to the length of one winding of the leading string L_1:

$$L_2 = L_1 = 2\pi r_1$$

2. The toryx eye radius r_0 is equal to a real positive constant:

$$r_0 = r_1 - r_2 = const.$$

3. The spiral velocity of the trailing string V_2 is constant and equals to the velocity of light in vacuum c at each point of its spiral path. Its components, the translational velocity V_{2t} and the rotational velocity V_{2r}, relate to the spiral velocity V_2 by the Pythagorean Theorem:

$$V_2 = \sqrt{V_{2t}^2 + V_{2r}^2} = c = const.$$

In spite of their apparent simplicity, the toryx spacetime postulates yield amazing spacetime properties of toryces. This is due to the fact that as the radius of the toryx leading string r_1 reduces from positive to negative infinity the toryx topology changes dramatically. At certain values of r_1 the toryx leading string, trailing string and spherical membrane become topologically-inverted inside out. In addition to that, either the translational velocity V_{2t} or the rotational velocity V_{2r} of trailing string becomes superluminal, while its spiral velocity V_2 remains equal to the velocity of light.

Thanks to these transformations, the toryces are capable to exist in four polarized states, creating the conditions for the formation of elementary matter particles by the unification of properly- matched polarized toryces. To provide a mathematical description of the toryx spacetime properties, it is necessary to modify several aspects of elementary math, including the definitions of zero, number line and elementary trigonometric functions. The modified elementary math is called the *spacetime math.* One of the most important features of the toryces that makes them different from many other proposed candidates for the prime elements of nature is related to their generic properties.

2. Generic Properties of the Toryces

The toryces are the smallest entities in the Universe that contribute to the self-preservation of the Universe thanks to their numerous properties found in many larger entities of the Universe. Take these properties away and the Universe, as we know it, will cease to exist.

Motion – The toryx is the smallest entity in the Universe with its components in the state of motion.

Propagation with the velocity of light – The toryx is the smallest entity in the Universe in which its trailing string propagates with velocity of light.

Limitation of degrees of spacetime freedom – The toryx is the smallest entity in the Universe with limited degrees of spacetime freedom.

The planetary motion – The toryx is the smallest entity in the Universe in which its leading strings follow the Spacetime Law of Planetary Motion for which the classical law of planetary motion is a particular case.

Spirality – The toryx is the smallest entity in the Universe involed in the spiral motion.

Polarization – The toryx is the smallest entity in the Universe that exists in the polarized states.

Expansion & contraction – The toryx is the smallest entity in the Universe that is capable to expand and contract.

Quantization – The toryx is the smallest entity in the Universe that exists in quantum spacetime (energy) states.

Absorption & release of spacetime – The toryx is the smallest entity in the Universe that is capable to absorb and release the spacetime (energy).

Unification & coexistence – The toryx is the smallest entity in the Universe that can be unified and coexist with an oppositely-polarized toryx, forming an elementary matter particle.

Radiation – The toryx is the smallest entity in the Universe that is capable to emit radiation in the form of a topologically-polarized helical spacetime called the *helyx*.

Crystallization – The toryces are the smallest entities in the Universe that create crystal structures to hold together elementary matter particles inside nucleons.

Spacetime properties – The toryces are the smallest entities in the Universe that possess clearly-defined spacetime properties, such as length, wavelength, radius, time, velocity and frequency.

Physical properties – The toryces are the smallest entities in the Universe that possess physical properties. The toryx physical properties are defined by their spacetime properties based on the following fundamental equation:

$$r_0 = \frac{e^2}{8\pi\varepsilon_0 m_e c^2}$$

where

e = elementary charge, ε_0 = electric constant, m_e = electron mass.

3. Metamorphoses of the Toryces

As the relative radius of the toryx leading string $b_1 = r_1/r_0$ decreases from positive to negative infinity, the steepness angle of the toryx trailing string φ_2 increases from 0 to 360^0.

Consequently, the toryx undergoes through the following four topological inversions (or turning inside out) as shown in Figure 3:

- Trailing string inverts at $\varphi_2 = 90^0$
- Spherical membrane inverts at $\varphi_2 = 180^0$
- Leading string inverts at $\varphi_2 = 270^0$
- All three toryx components invert at $\varphi_2 = 0/360^0$.

4. Excitation and Oscillation of the Toryx

Toryces exist in excitation and oscillation quantum states as shown in Figure 4. During the excitation of a toryx its eye radius r_0 remains constant while other parameters change in quantum steps as a function of the *toryx quantization parameter* z given by the equations:

$$b_1 = z = 2(n\Lambda)^m$$

where:

m = 0, 1, 2, … toryx exponential excitation quantum state

n = 0, 1, 2, … toryx linear excitation quantum state

Λ = 137, *spacetime quantization constant.*

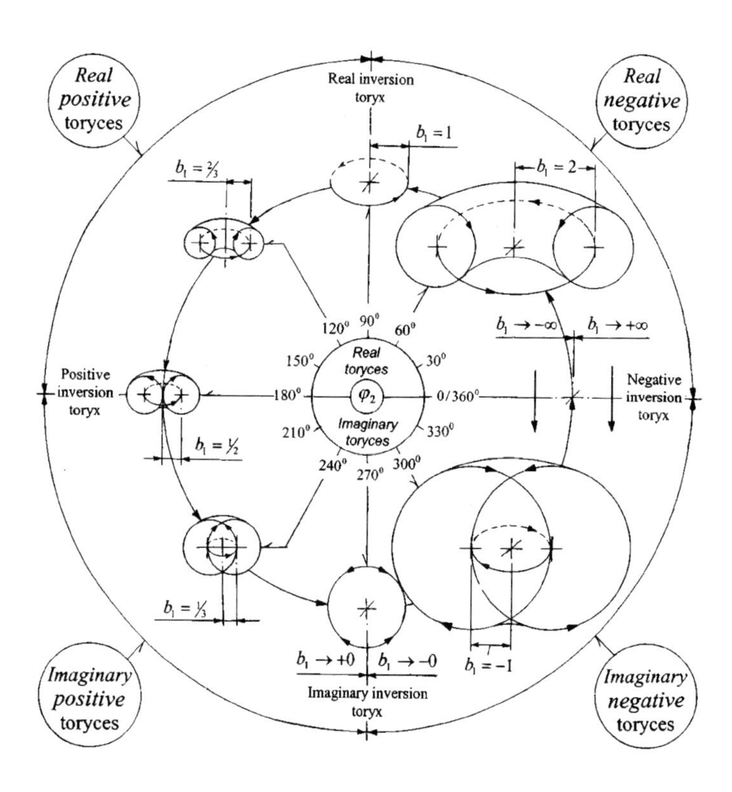

Figure 3. Metamorphoses of the toryx leading and trailing strings as a function of the steepness angle of the trailing string φ_2 (spherical membranes are not shown).

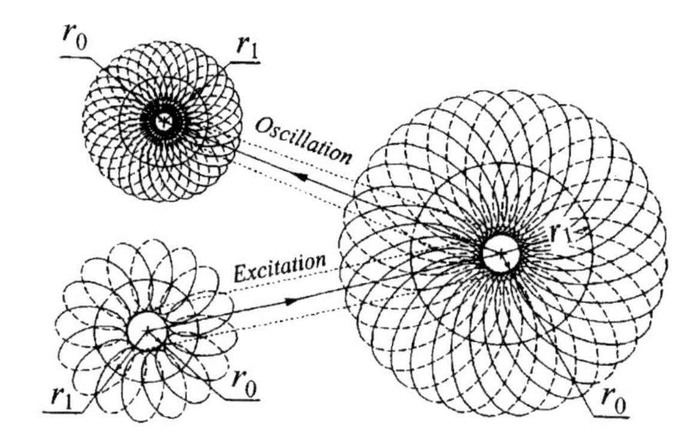

Figure 4. Excitation and oscillation of a toryx.

The exponential excitation quantum state m of the toryces depends on the *spacetime levels*: *L1* (dark energy), *L2* (ordinary matter), *L3* (dark matter), etc. The linear excitation quantum states n exist within each spacetime level. During its oscillation the entire toryx, including its eye radius r_0, reduces in size and its all spacetime parameters change in the quantum steps defined by the *toryx oscillation quantum states* q = 0, 1,…

5. Elementary Matter Particles

Four kinds of elementary matter particles are formed by the unification of the topologically-polarized toryces: *electrons, positrons, ethertrons* and *singulatrons* as shown in Figure 5. Consequently, the toryx undergoes through the following four topological inversions (or turning inside out) as shown in Figure 3:

- Trailing string inverts at $\varphi_2 = 90^0$
- Spherical membrane inverts at $\varphi_2 = 180^0$
- Leading string inverts at $\varphi_2 = 270^0$
- All three toryx components invert at $\varphi_2 = 0/360^0$.

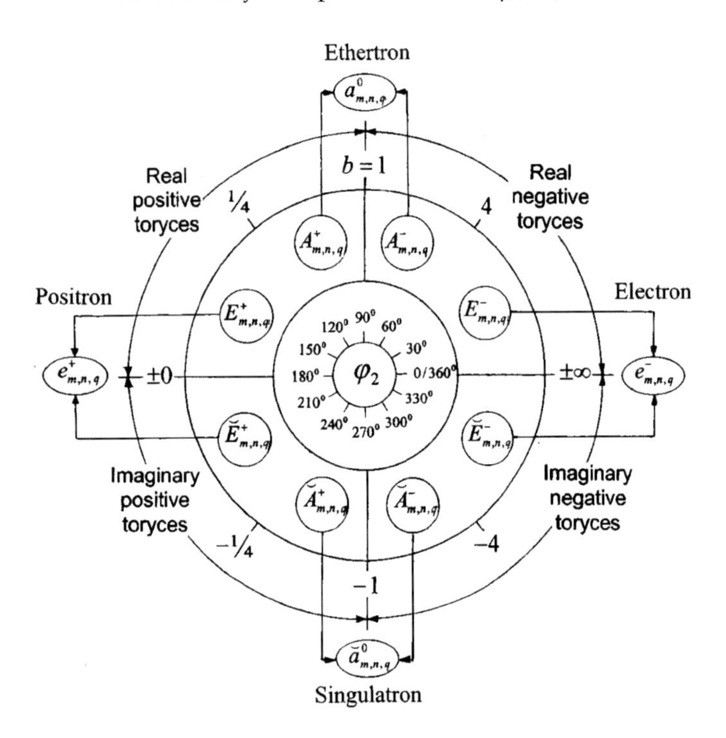

Figure 5. Formation of four elementary matter particles from polarized toryces.

Stable elementary matter particles are formed only if they follow the proposed *Spacetime Conservation Law* according to which the sum of polarization parameters of their constituent toryces must be infinitely small.

7. Metamorphoses of the Helyces

Similarly to the toryx, the degrees of freedom of helyces are also limited, allowing them to form four kinds of topologically-polarized helyces shown in Figure 8.

6. Helyx Structure

Excited and oscillated toryces emit spiral spacetimes called the *helyces*. The helyx has two levels. As shown in Figure 6, the first level is in the form of a double-helical spiral forming the helyx *leading string* A_1. The second level of the helyx is made up of two double helices forming the helyx *trailing string* as shown in Figure 7. Four branches of the helyx trailing string A_2 reside inside a *cylindrical membrane*; they are wrapped around two branches of the helyx leading string A_1.

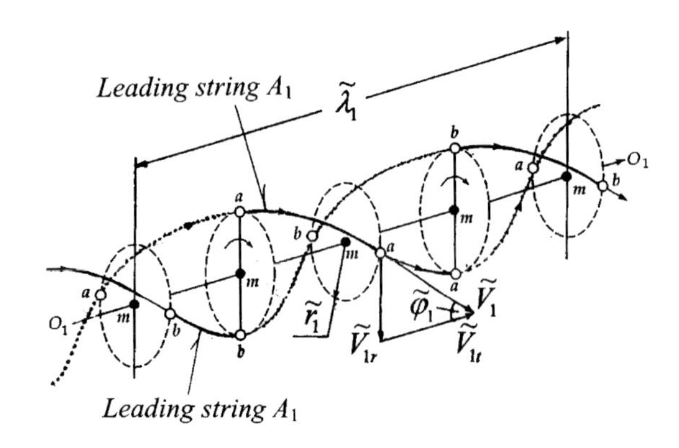

Figure 6. Structure of the helyx leading string.

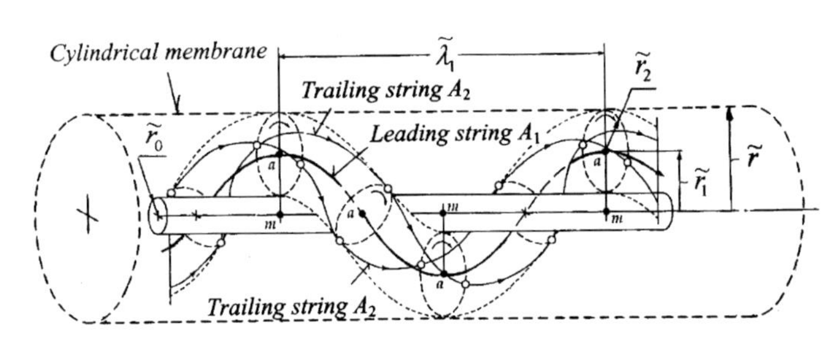

Figure 7. One of two double-helical leading strings A_1 accompanied by two branches of the trailing string A_2 residing inside a cylindrical membrane.

8. Elementary Radiation Particles

Four kinds of elementary radiation particles are formed by the unification of the topologically-polarized helyces: *electrons, positons, ethertons* and *singulatons* as shown in Figure 9. The names of the elementary radiation particles are similar to the names of their parental elementary matter particles responsible for the creation of helyces. Singulatons propagate with superluminal velocities, while electons, positons and ethertons propagate with velocity of light.

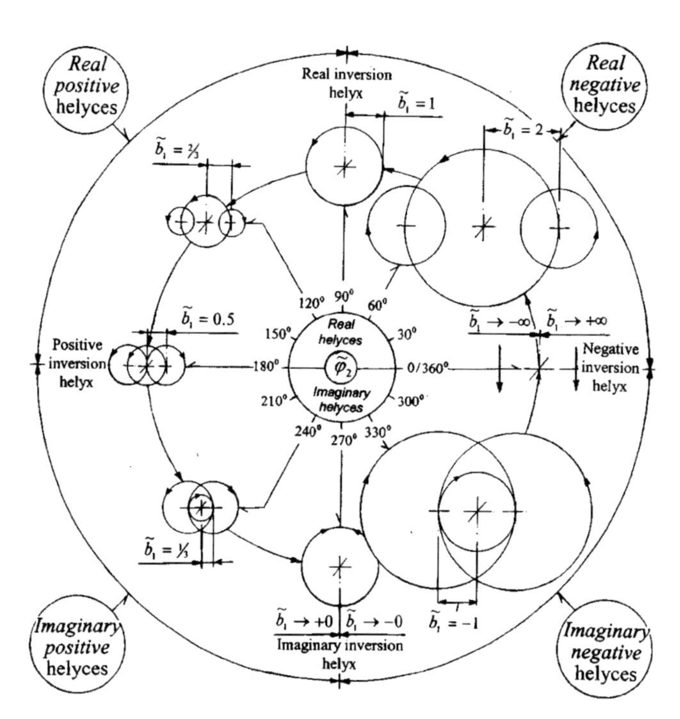

Figure 8. Transformations of cross-sections of helyces as a function of the apex angle of the trailing string φ_2.

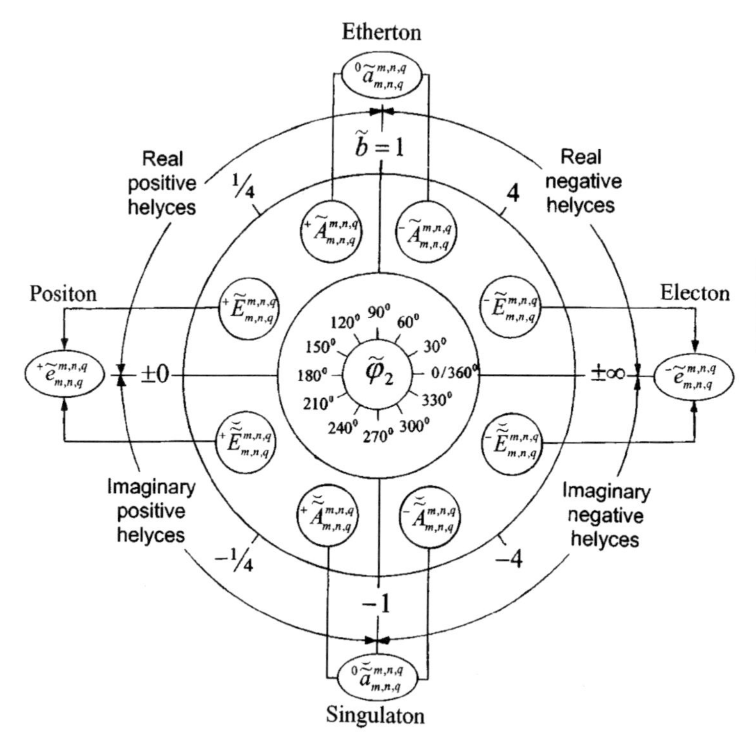

Figure 9. Formation of four elementary radiation particles from polarized helyces.

9. Spacetime Levels of the Universe

According to the UST, the universe exists at several spacetime levels L. The properties of the first three spacetime levels are shown in the Tables 1 and 2.

- At each spacetime level, the properties of elementary matter particles depend on the exponential excitation quantum states m of their constituent toryces.
- Each spacetime level of the Universe is identified by the frequency ranges and velocities of the radiations emitted by their elementary matter particles.
- The radiations emitted by the electrons, positrons and ethertrons propagate with the velocity of light, while the radiations emitted by the singulatrons propagate with superluminal velocities.
- As the matter levels increase:

  a) The masses of nucleons increase and the frequencies of radiations emitted by the nucleons increase.

  b) The orbital radii of atomic electrons increase and the frequencies emitted by the atomic electrons decrease.

  c) The densities of atoms decrease.

Table 1. Properties of the first three spacetime levels of the Universe.

Spacetime levels L	Frequencies & velocities of radiations emitted by			
	Ethertron	Singulatron	Electron	Positron
$L1$	$m = 0$ luminal	$m = 0$ superluminal	$m = 1$ luminal	$m = 1$ luminal
$L2$ (ordinary matter)	$m = 1$ luminal	$m = 1$ superluminal	$m = 2$ luminal	$m = 2$ luminal
$L3$	$m = 2$ luminal	$m = 2$ superluminal	$m = 3$ luminal	$m = 3$ luminal

Table 2. Relative properties of the hydrogen atoms $\downarrow H_L^0$ of several spacetime levels L.

Hydrogen atoms	Electron orbital radius ratio	Orbital magnetic moment ratio	Mass ratio
$\downarrow H_{L1}^0$	1/137.2	1/12	1/107
$\downarrow H_{L2}^0$ (ordinary matter)	**1.0**	**1.0**	**1.0**
$\downarrow H_{L3}^0$	137.0	11.72	136.7

10. Definition of Dark Matter & Dark Energy

The UST provides the following definition of the dark matter and dark energy.

Dark matter - According to the UST, the "ordinary matter" is related to the spacetime level *L2* which atomic electrons emit the radiations within the frequency ranges of observed *infrared, visible* and *ultraviolet cosmic radiations*. The "dark matter" is related to all other spacetime levels *L1, L3, L4*, etc..

Here is how the properties of the dark matter compare with the properties of the ordinary matter:

- In the dark matter of the spacetime level *L1*, the nucleons are about 107 lighter than in the ordinary matter, the orbital radii of their atomic electrons are about 137 times smaller and they emit the radiations within the frequency range of observed *cosmic X-ray background (CXB) radiation*.
- In the dark matter of the spacetime level *L3*, its nucleons are about 137 times heavier than in the ordinary matter, the orbital radii of their atomic electrons are about 137 times larger and they emit the radiations within the frequency range of observed *cosmic microwave background (CMB) radiation*.
- As the spacetime levels *L* of the dark matter continues to increase, its nucleons become progressively heavier, the orbital radii of their atomic electrons become larger and they emit the radiations within the frequency range of the observed *cosmic radio wave radiation* and lower.

Dark energy - According to the UST, the apparent expansion of the universe is not due to the global motion of galaxies apart from one another. This phenomenon is caused by the local expansions of the spacetimes from lower to higher levels *L*.

11. Additional Predictions of the UST

The UST predicts the following already known information:

- Orbital radii of atomic electrons of hydrogen atom
- Orbital velocities of the atomic electrons of hydrogen atom
- Mass, charge and magnetic moment of proton
- Mass, charge and magnetic moment of neutron
- Mass, charge and magnetic moment of muon
- Mass and charge of tau particle
- Decay of a neutron into a proton, electron and electron neutrino.

The UST makes the following new predictions:

- There is a stable neutron.
- Leptons are merely the oscillated excited electrons.
- Neutrinos are merely the radiation particles emitted by the oscillated excited electrons.

The UST provides two possible predictions regarding to the nature of new particles detected by the CERN Large Hadron Collider (LHC) experiments reported in 2012:

1. After their collisions, the excited protons are transferred from the ordinary spacetime level *L2* to the dark spacetime level *L3*. The new radiation particles (bosons) are emitted by the de-excited protons transferred back from the dark spacetime level *L3* to the ordinary spacetime level *L2*.
2. The new particles are the excited protons that after their collisions are transferred from the ordinary spacetime level *L2* to the dark spacetime level *L3* and then captured electrons.

12. Macro-Trons & Macro-Tons

Macro-toryces are associated with *celestial bodies* (Fig. 10), and they are responsible for the interactions between these bodies. According to the UST, a celestial body is defined as an entity made up of elementary matter particles.

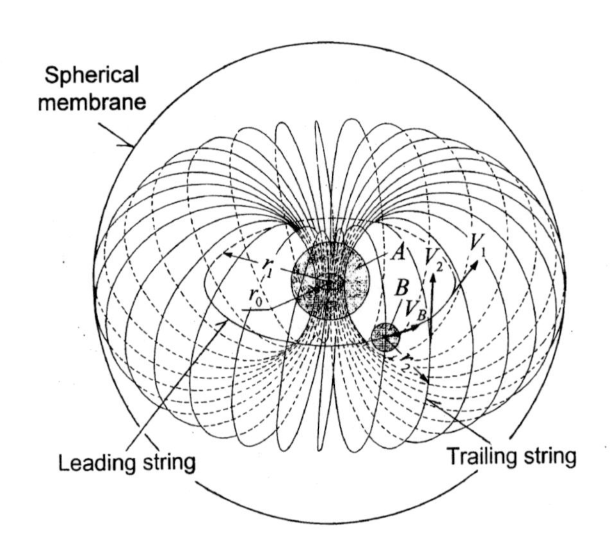

Figure 10. Central body *A* with a macro-toryx encompassing the satellite body *B*.

Spacetime properties of the macro-trons are described by the same equations, except for the toryx eye radius r_0 that is dependent on the mass of each celestial body. Similarly to the four kinds of micro-toryces shown in Figure 3, there are four kinds of topologically-polarized *macro-toryces* that exist in excitation quantum states. Similarly to the four kinds of micro-trons shown in Figure 5, there are four kinds of *macro-trons* formed by the unification of the topologically-polarized macro-toryces. Macro-trons are associated with various celestial bodies, including black holes, and they are responsible for the interactions between celestial bodies. Similarly to the micro-trons, the macro-trons emit the *macro-tons* made up of the topologically-polarized *macro-helyces*.

The UST explains gravity as a result of violation of the derived Spacetime Law of Planetary Motion. Consequently, the UST derives the Spacetime Law of Gravitation for which Newton's universal law of gravitation is a particular case.

References

[1] Ginzburg, V.B., "Toroidal Spiral Field Theory," *Speculations in Science and Technology*, Vol. 19, 1996.

[2] Ginzburg, V.B., *Spiral Grain of the Universe - In Search of the Archimedes File*, University Editions, Inc., Huntington, WV, 1996.

[3] Ginzburg, V.B., "Structure of Atoms and Fields," *Speculations in Science and Technology*, Vol. 20, 1997.

[4] Ginzburg, V.B., "Double Helical and Double Toroidal Spiral Fields," *Speculations in Science and Technology*, Vol. 22, 1998.

[5] Ginzburg, V.B., *Unified Spiral Field and Matter - A Story of a Great Discovery*, Helicola Press, Pittsburgh, PA, 1999.

[6] Ginzburg, V.B., "Nuclear Implosion," *Journal of New Energy*, Vol. 3, No. 4, 1999.

[7] Ginzburg, V.B., *Continuous Spiral Motion System for Rolling Mills*, U.S. Patent No. 5,970,771, Oct. 26, 1999.

[8] Ginzburg, V.B., *Continuous Spiral Motion System and Roll Bending System for Rolling Mills*, U.S. Patent No. 6,029,491, Feb. 29, 2000.

[9] Ginzburg, V.B., "Dynamic Aether," *Journal of New Energy*, Vol. 6, No. 1, 2001.

[10] Ginzburg, V.B., *The Unification of Strong, Gravitational & Electric Forces*, Helicola Press, Pittsburgh, PA, 2003.

[11] Ginzburg, V.B., "Electric Nature of Strong Interactions," *Journal of New Energy*, Vol. 7, No. 1, 2003.

[12] Ginzburg, V.B., "Unified Spiral Field Theory – A Quiet Revolution in Physics," *VIA-Vision in Action*, Vol. 2, No. 1 & 2, 2004.

[13] Ginzburg, V.B., "The Relativistic Torus and Helix as the Prime Elements of Nature," *Proceedings of the Natural Philosophy Alliance*, Vol. 1, No. 1, Spring 2004.

[14] Ginzburg, V.B., *Prime Elements of Ordinary Matter, Dark Matter & Dark Energy*, Helicola Press, Pittsburgh, PA, 2006.

[15] Ginzburg, V.B., *Prime Elements of Ordinary Matter, Dark Matter & Dark Energy – Beyond Standard Model & String Theory*, The second revised edition, Universal Publishers, Boca Raton, Florida, 2007.

[16] Ginzburg, V.B., "The Unification of Forces," *Proceedings of the Natural Philosophy Alliance*, Vol. 4, No. 1, 2007.

[17] Ginzburg, V.B., "The Origin of the Universe, Part 1: Toryces," *Proceedings of the Natural Philosophy Alliance*, The 17th Annual Conference of the NPA 23-26 June 2010 at California State University Long Beach, Vol. 7.

[18] Ginzburg, V.B., "Basic Concept of 3-Dimensional Spiral String Theory (3D-SST)," *Proceedings of the Natural Philosophy Alliance*, The 18th Annual Conference of the NPA, 6-9 July 2011 at the University of Maryland, College Park, USA, Vol. 8.

[19] Ginzburg, V.B., *The Spacetime Origin of the Universe*, Helicola Press, Pittsburgh, PA, 2013.

[20] Ginzburg, V.B., *The Spacetime Origin of the Universe with Visible Dark Matter & Energy*, Second Edition, Helicola Press, Pittsburgh, PA, 2015.

The Correct Derivation of Kepler's Third Law for Circular Orbits Reveals a Fatal Flaw in General Relativity Theory

Jaroslav Hynecek [1]
[1] Isetex, Inc., 905 Pampa Drive, Allen, TX 75013, USA, © 6/15/2015
Correspondence: Jaroslav Hynecek, Isetex, Inc., 905 Pampa Drive, Allen, TX 75013, USA.
E-mail: jhynecek@netscape.net

In this paper the Kepler's third law is derived for circular orbits using the two different metrics. The resulting formulas are compared with the formula for the Kepler's third law derived from the Newtonian physics. The derivation is using the Lagrange formalism, but comments are made on error in derivation that has appeared in previous publication. It is found that the Kepler's third law derived using the Schwarzschild metric results in an identical formula obtained from the Newtonian physics of a flat spacetime geometry. This clearly illustrates a problem for the Schwarzschild metric and consequently for the General Relativity Theory.

Key words: Lagrange formalism, Kepler's third law, relativistic Kepler's third law, Schwarzschild metric, metric derived in the Metric theory of gravity, errors in the General Relativity Theory

1. Introduction

The Kepler's third law is a very important law for astronomers, which is used to determine the mass of planets and stars based on the gravitational constant measured here on Earth and on the time of the planet's full orbit completion. The discovery of this law played an important role in the past in advancing the knowledge about our Solar system neighborhood and in convincing astronomers that the planets orbit the Sun and that the Moon orbits Earth. Because the time of the orbits can be measured with a high precision and the radius of the orbits is also reasonably well known, the mass of the centrally gravitating bodies can thus be found very accurately.

This law is easily derivable from the Newton inertial and gravitational laws for a circular orbit by equating the inertial centrifugal force with the gravitational force as follows:

$$\frac{v^2}{r} = \frac{\kappa M}{r^2} \tag{1}$$

where M is the mass of the centrally gravitating body and κ the gravitational constant. A hidden assumption used in this formula derivation is the absolute equality of inertial and gravitational masses, which is not strictly true [1]. The correction for this effect is given in section **4**. By realizing that the average velocity is the length of the orbit circumference divided by the time of the orbit completion Equation 1 can be rewritten in the familiar third Kepler's law form:

$$t_{nt}^2 = \frac{4\pi^2 r^3}{\kappa M} \tag{2}$$

With many advances in the theory of gravity from the Newtonian to Einstein's General Relativity Theory (GRT) and further to more general Metric Theories of Gravity (MTG) it is thus natural to ask how is this law changed and is it accurate enough to determine, for example, the mass of our Sun with enough precision so that no large trajectory errors are generated when the space probes are sent to investigate other planets of our Solar system.

It is fascinating to see that this law plays again an important role in showing that the GRT is not the correct theory of gravity, similarly as the old planetary epicycle theory was shown incorrect, and that the GRT thus needs to be fundamentally changed.

2. The derivation of Kepler's third law for a general spacetime metric of a centrally gravitating mass

Several derivations of this law have been already published in the literature [2, 3]. The references given here are for the comparison purposes of various assumptions used in GRT derivations and in MTG derivation and the conclusions obtained from them. The derivation presented in this paper is very basic and more importantly it does not depend on the validity of the GRT. The Kepler's third law for a motion of a small test body will therefore be derived first in a general form and then applied to two key cases: the Schwarzschild metric spacetime and the new metric spacetime derived previously by the author [1]. It is, of course, possible to apply the derived formulas to other metrics that can be found published in the literature, but the new metric satisfies the same four observational tests of GRT for the weak gravitational fields as the Schwarzschild metric does so it is interesting to make a comparison only between these two.

The general differential metric line element of a spacetime of a non-rotating centrally gravitating body is as follows:

$$ds^2 = g_{tt}(cdt)^2 - g_{rr}dr^2 - g_{\varphi\varphi}d\Omega^2 \tag{3}$$

where: c is the local intergalactic speed of light, $d\Omega^2 = d\vartheta^2 + \sin^2\vartheta d\varphi^2$, $g_{tt} = \exp(2\varphi_v)$, $g_{tt}\, g_{rr} = 1$, and where the metric coefficients depend only on the radial coordinate. This form of metric assumes that according to the Riemann hypothesis the motion can be represented by a curved spacetime in which the small test bodies move in a free fall along geodesic lines and are not experiencing any forces in contrast to a flat spacetime with fields and forces that guide the motion. This concept forms the basis for all MTG theories and has also been adapted by Einstein in his derivation of general relativity. The Einstein's GRT, however, includes additional assumptions related to the Ricci tensor that led to the derivation of Einstein field equations with the Schwarzschild metric as a solution. The Riemann principle is thus more general in comparison to the GRT and allows derivation of other metrics describing the spacetime not only the Schwarzschild metric. In the new metric derived previously [1, 5] the metric coefficients are: $g_{tt} = \exp(-R_s/\rho)$, $g_{tt}\, g_{rr} = 1$, and $g_{\varphi\varphi} = \rho^2 g_{tt}$, while for the Schwarzschild metric they are: $g_{tt} = (1-R_s/r)$, $g_{tt}\, g_{rr} = 1$, and $g_{\varphi\varphi} = r^2$. The Schwarzschild radius R_s is defined as usual as follows: $R_s = 2\kappa M/c^2$. Using the well-known and ages tested Lagrange formalism, considering for simplicity motion only in the equatorial plane, the Lagrangian describing such motion of a small test body in this spacetime is then as follows:

$$L = g_{tt}\left(\frac{cdt}{d\tau}\right)^2 - g_{rr}\left(\frac{dr}{d\tau}\right)^2 - g_{\varphi\varphi}\left(\frac{d\varphi}{d\tau}\right)^2 \tag{4}$$

The first integrals of Euler-Lagrange (EL) equations corresponding to the time and angle coordinates derived from the variational principle $\delta\int_\tau L d\tau = 0$ are thus:

$$g_{tt}\left(\frac{dt}{d\tau}\right) = k \qquad g_{\varphi\varphi}\left(\frac{d\varphi}{d\tau}\right) = \alpha \tag{5}$$

where k and α are arbitrary constants of integration. The EL equation of motion corresponding to the radial coordinate is as follows:

$$-\frac{d}{d\tau}\left(2g_{rr}\frac{dr}{d\tau}\right) = \dot{g}_{tt}\left(\frac{cdt}{d\tau}\right)^2 - \dot{g}_{rr}\left(\frac{dr}{d\tau}\right)^2 - \dot{g}_{\varphi\varphi}\left(\frac{d\varphi}{d\tau}\right)^2 \tag{6}$$

where the dot represents the partial derivative with respect to the radial coordinate. Since for the circular orbits the radial coordinate is constant with: $dr/d\tau \Rightarrow 0$, and $d^2r/d\tau^2 \Rightarrow 0$, Equation 6 simplifies to read:

$$\left(\frac{d\varphi}{dt}\right)^2 = c^2\frac{\dot{g}_{tt}}{\dot{g}_{\varphi\varphi}} \tag{7}$$

In this formula the first integral corresponding to the time coordinate shown in Equation 5 was used to eliminate the non-observable variable τ. Considering now that the coordinate orbital time t_o, which is the observable quantity referenced to the central mass coordinate system, is found when the angle is set to: $\varphi = 2\pi$, the following equation is obtained:

$$t_o = \frac{2\pi}{c}\sqrt{\frac{\dot{g}_{\varphi\varphi}}{\dot{g}_{tt}}} \tag{8}$$

This is the general formula that can be used for any metric describing the spacetime of a non-rotating centrally gravitating body that conforms to a form given in Equation 3. For the Schwarzschild metric the result is:

$$t_{os} = \frac{2\pi}{c}\sqrt{\frac{2r^3}{R_s}} = 2\pi\sqrt{\frac{r^3}{\kappa M}} \tag{9}$$

Surprisingly this result is identical with the Newtonian case derived in Equation 2, which indicates that the Schwarzschild metric spacetime with the Ricci curvature tensor equal to zero does not have any effect on the planetary orbital period. Apparently even the event horizon does not seem to pose any problems for the orbital time. For example, inside of the Black Hole (BH) at $r = R_s/2$ the test bodies should whiz around at the vacuum speed of light and at the smaller radius even faster. This does not seem reasonable and therefore this metric does not describe the reality correctly as already discussed elsewhere [4, 5]. For the new metric, however, the result is:

$$t_{oh} = 2\pi\sqrt{\frac{\rho^3}{\kappa M} + \frac{\rho^2}{c^2}} \tag{10}$$

where the physical distance $\rho = \rho(r)$ is a function of the natural coordinate distance r and is calculated using the following differential equation obtained from the metric:

$$d\rho = e^{R_s/2\rho}dr \tag{11}$$

For more clarity in understanding of these differences in orbital time formulas the results are for convenience summarized in a table below:

Spacetime type	Metrics/Formula	Orbit time formula
Flat	Newton-Kepler $m_g = m_i$	$t_{nt} = 2\pi\sqrt{\frac{2r^3}{c^2R_s}}$
Curved	Schwarzschild metric	$t_{os} = t_{nt} = 2\pi\sqrt{\frac{2r^3}{c^2R_s}}$
	New metric	$t_{oh} = 2\pi\sqrt{\frac{2\rho^3}{c^2R_s} + \frac{\rho^2}{c^2}}$

The result in Equation 10 indicates that the orbital time has a limit for large M equal to the physical length of the path divided by the speed of light. This is reasonable and easily understandable for the new metric since the orbital motion cannot exceed the speed of light. For the Schwarzschild metric formula, however, there is no such limit, which presents a significant problem for this metric and consequently for the GRT. The Schwarzschild metric describes the reality only approximately and should not be used to model the spacetimes with strong gravitational fields.

3. The derivation errors that are often made

In order to shorten the calculations it may be tempting to simplify the above presented derivation procedure and use the Lagrangian itself as the first integral. The fact that the Lagrangian is also a first integral equal to: $L = c^2$ can be found proven in many publications. The computation using this first integral and the first integral for the time coordinate with $k = 1$ as shown in Equation 5 would thus proceed as follows:

$$L_0 = g_{tt}\left(\frac{cdt}{d\tau}\right)^2 - g_{\varphi\varphi}\left(\frac{d\varphi}{d\tau}\right)^2 \quad (12)$$

resulting in the formula:

$$t_0 = \frac{2\pi}{c}\sqrt{\frac{g_{\varphi\varphi}}{g_{tt}(1-g_{tt})}} \quad (13)$$

which is markedly different from the correct formula shown in Equation 10. The approach similar to this one was used by Hynecek [6] and it is unfortunately incorrect. The correct calculation is available in another publication by Hynecek [7], but this publication is not easily accessible and for this reason it is repeated here. The derivation error results from an incorrect imposition of constraint $dr/d\tau \to 0$ on the Lagrangian in Equation 4 before the variations are carried out. The Lagrangian $L_0 = c^2$ in Equation 12 is, therefore, not the correct Lagrangian and consequently results in an incorrect first integral.

4. Correction of standard Newton-Kepler formula for the different gravitational and inertial mass dependencies on velocity

It is generally believed by the mainstream relativistic physicists that the inertial mass and the gravitational mass of a test body depend on velocity the same way. This is sometimes called the Einstein weak equivalence principle (WEP). However this author has previously shown that this is not true and that the following dependencies for the inertial and gravitational masses on velocity hold [1]:

$$m_i = \frac{m_0}{\sqrt{1-v^2/c^2}} \quad (14)$$

This is the standard formula of Special Relativity Theory (SRT) for the inertial mass dependency on velocity with m_0 being the rest mass. For the gravitational mass, however, the dependency on velocity is as follows:

$$m_g = m_0\sqrt{1-v^2/c^2} \quad (15)$$

This is a new formula that cannot be found anywhere in the standard published literature. Substituting these two formulas into Equation 1 then results in the following:

$$\frac{v^2}{r} = \frac{\kappa M}{r^2}\left(1-\frac{v^2}{c^2}\right) \quad (16)$$

After some simple algebra rearrangements, using again the relations for velocity, the time of the orbit completion, and the length of the orbit, the corrected Kepler's third law formula for circular orbits becomes equal to:

$$t_{ntc} = 2\pi\sqrt{\frac{r^3}{\kappa M} + \frac{r^2}{c^2}} \quad (17)$$

This is a nice consistency check for the formula derived in Equation 10, since it is possible to assume that for large distances the spacetime is almost Minkowski flat resulting in an approximate validity of relation: $\rho(r) \sim r$. The Kepler's third law thus not only confirms the correctness of the new MTG metric, but also the correctness of Equation 15.

Another consistency check can be made by letting the speed of light in Equation 17 tend to infinity. This typically transforms the relativistic formulas to the standard Newtonian physics formulas and, therefore, also transforms the formula in Equation 17 to the classical formula of Equation 2.

The gravitational mass dependency on velocity, as shown in Equation 15 is, unfortunately, not recognized by the relativistic mainstream scientists, but it has far reaching consequences. It invalidates the GRT with its Black Holes and the Big Bang theory that relies on the absolute equivalence of these two masses and at the same time confirms that photons do not have a gravitational mass; they have only an inertial mass. Photons are, therefore, not attracted by the gravitating bodies, they move along the geodesic (generalized straight) lines, and their curved paths in the Universe clearly show that the spacetime is a material entity, the dark matter, which is deformed by the gravity of massive bodies.

5. The dire consequences for the GRT

The fact that the correct derivation of the GRT Kepler's third law leads to the same formula as the formula derived from the Newtonian physics of flat spacetime geometry is well known to many mainstream relativists. They can even derive the Schwarzschild metric from the Kepler's third law [8]. The typical excuse that is often used is that this is due to the lucky choice of coordinates. This, of course, cannot be true. The formula presented in Equation 10 is an invariant. It does not matter what coordinates are used, because the physical coordinates are always the same and are not affected by the gravity. The formula also clearly includes the curvature of spacetime, which is described by the relationship between the physical coordinate ρ and the natural coordinate r. The curvature of spacetime is clearly not included in Equation 9, which is an obvious and glaring problem. It is thus clear that the Schwarzschild metric does not correspond to reality and consequently the GRT is the incorrect theory of gravity and should be abandoned.

It is fascinating to see that the Kepler's third law for the circular orbits dispels again the myths of "relativistic epicycles" that the mainstream relativists so desperately adhere to. Unfortunately the facts do not matter here; it is the religion of GRT and its ideology that is not permitted to be challenged.

6. Conclusions

In this article it was clearly shown that the Schwarzschild metric of GRT does not correspond to reality. This is a fatal problem for the theory. The derivation used the Kepler's third law for circular orbits to show this problem. It was also shown that the previously published derivation contains a subtle error. The error was analyzed and its origin clearly explained. It is thus necessary to always correctly use and correctly adhere to the well tested and proven Lagrange formalism that so beautifully describes the physics of curved spacetimes as it was introduced by Riemann and others.

It is also necessary to mention that this derivation did not take into account the repulsive dark matter [9] that permeates all the space and manifests itself when cosmological distances are involved. The Kepler's third law will thus have to be modified for such distances.

One can only wonder when the mainstream relativistic scientists will realize this problem and abandon the GRT with its preposterous Black Holes and Big Bang theory. Perhaps this will take another 100 years before enough conflicting data and observations accumulate and the theory crumbles under the weight of this evidence [10, 11]. Unfortunately this will not happen during my lifetime, so I will not be able to enjoy this wonderful paradigm shift. For the time being I am only enjoying the discovery of a very small, but beautiful, piece of eternal truth. Similar papers criticizing the GRT can easily be accessed elsewhere [12, 13].

References

1. J. Hynecek, "Remarks on the Equivalence of Inertial and Gravitational Masses and on the Accuracy of Einstein's Theory of Gravity", Physics Essays volume 18, number 2, 2005.
2. R. Y. Kezerashvili and J. F. V´azquez-Poritz, "Deviations from Keplerian Orbits for Solar Sails", arXiv:0907.3311v1 [gr-qc] 20 July 2009.
3. B. M. Barker and G. G. Byrd and R. F. O'Connel, "Relativistic Kepler's Third Law", The Astrophysical Journal, 305, pp. 623-633, June 15 1986.
4. J. Hynecek, "The Galileo effect and the general relativity theory", Physics Essays, volume 22, number 4, 2009.
5. J. Hynecek, "Can the Geometry Prove the General Relativity Incorrect?" http://vixra.org/abs/1408.0053.
6. J. Hynecek, "Relativistic Third Kepler Law for Circular Orbits", http://www.gsjournal.net/Science-Journals/%7B$cat_name%7D/View/1505.
7. J. Hynecek, "Kepler's Third Law for Circular Orbits Derived in Metric Theory of Gravity", Physics Essays, volume 23, number 3, 2010, pp. 502-505.
8. http://mathpages.com/rr/s5-05/5-05.htm.
9. J. Hynecek, "The Repulsive Dark Matter Model of the Universe Relates the Hubble Constant to the CMBR Temperature" http://www.ccsenet.org/journal/index.php/apr/article/viewFile/21114/15986
10. Quotation From The Daily Galaxy: March 15, 2014 Fifteen Old, Massive Galaxies Found in the Early Universe --"They Shouldn't Even Exist".
11. 'Methuselah', a 14.5 billion years old star: http://www.nasa.gov/mission_pages/hubble/science/hd140283.html
12. http://vixra.org/author/jaroslav_hynecek
13. http://www.gsjournal.net/Science-Journals-Papers/Author/201/Jaroslav,%20Hynecek

Fusion Mass Losses and Tunnels Formed between Touching Nucleons

Carl R. Littmann
Washington Lane, Apt. 313, Wyncote, PA 19095
e-mail: clittmann@verizon.net

When nucleons fuse with neighbors, there is mass loss and great energy emitted. We discuss the great jump in fusion mass lost when four nucleons are fused together compared to three. About 4 times as much mass is lost, but we show how this is largely expected since 4 times as many triangular planes with 'donut holes' are formed vs. for 3 nucleons. Specifically, we show how after 3 Hydrogen-1 atoms fuse to form 1 Helium-3 atom, the mass lost equals twice the mass of a sphere sized to barely fit through the array's donut hole, with an error of about 1 part in 5000. Finally, we discuss implications of all the above and related topics, including supplementing the present neutron-proton based 'binding energy' method with a more revealing electron-proton based one.

1. Introduction

1.1. The Main Motivations for the Paper

To prevent confusion, readers should remember that this paper actually addresses two different fusion subjects.

The first is about trying to understand the reasons for the very different amounts of mass lost during different fusion events, that is when two, three or four nucleons are fused together. And we try to predict those different results with satisfactory accuracy. Regarding that, we note that when 3 nucleons touch, only 1 triangular structure is formed with 1 hole in it, like a donut hole. But in the 4-nucleon case, we have 4 triangular planes and 4 holes.

We hypothesize that those holes serve as tunnels, and invite a flowing action through those tunnels. That flow results in a decrease in the relative pressure between nucleons, a 'Venturi suction' or attraction-like effect. And that is related to Bernoulli's equation. So such major tunnels appear when three or more nucleons touch, and thus those nucleons cling together especially strongly. Therefore, nuclei can be made consisting of more than just one nucleon or two weakly bound ones. And that makes our amazingly diverse world possible! That is indicated in Fig. 1 and Fig. 2.

Where appropriate, for some calculations in this paper we will use some aspects of Bohr's 'liquid drop model' of the nucleus, for example, incompressibility and uniform density.

The other major subject in this paper is this: There is an old method of processing information related to nuclear 'binding energies' that is barely adequate for noting some of the relationships discovered in this paper. And that old method has some flaws. To fully appreciate the closeness of relationships discussed in this paper, and to correct problems inherent in the old binding energies methodology, we will discuss a better one.

Important: The common old binding energy method will be designated here as the "n,p-based binding energy method" -- because it is based on imagining the neutron and proton as the basic building blocks. But our newer advocated method will be designated as the "e,p-based binding energy method" -- because it is based on imagining the electron and proton as the basic building blocks of nuclei. Actually, since the 'Bohr Atom' employs one electron orbiting around one proton, the e,p-based binding energy method is equivalent to using one or more Bohr Atoms as building blocks. In fact, historically, that system has been previously and successfully used by the famous textbook writers and accomplished scientists, Sears and Zemansky [1]. I.e., even before this present paper.

A major obvious failure of the old n,p-based binding energy method arises when it is used to calculate binding energies for ${}_1H^3$ and ${}_2He^3$, (Hydrogen-3 and Helium-3 isotopes). That old method calculates Hydrogen-3 as having much more binding energy than stable Helium-3. But sadly Hydrogen-3 turns out to be unstable and decays. That misleading calculation occurs because the old binding energy method generally uses the unstable, complex neutron as one of the basic building blocks, instead of the more basic and stable electron mass and proton mass.

But using our e,p-based binding energy method, the calculations show Helium-3 with slightly more binding energy than Hydrogen-3. That's good and appropriate because Helium-3 is stable. More details about the calculations are given later.

(Ideally, I wish a different terminology, instead of 'binding energy', had been used from the very beginnings of nuclear science, but we'll leave that alone for now, and return to our first major topic.)

One of the most striking things noticed when one studies physics and nuclear fusion is the huge energy that is emitted when four nucleons fuse together to form the common helium nucleus, (${}_2He^4$). Even with only a general science background, much of the public will associate that fusion energy with the powering of the Sun and likely a hydrogen bomb.

Our solar system's mass consists of about 75% hydrogen and 25% helium and less than 1% other nuclei. So we are especially motivated to focus on fusion involving 4 nucleons and fewer nucleons. In fact, in terms of numbers of nuclei (not 'weight'), hydrogen's proton nucleus alone constitutes about 94% of the nuclei in our solar system.

With somewhat more study, one notes the huge jump in emitted energy when four nucleons are fused together vs. only three. That is the approximate quadrupling of output as described in the Abstract above, and is even evident from old common graphs of n,p-based binding energy per nucleon.

That quadrupling is not the expected outcome based on most people's experience with other things. For example, if one touches 2 sticky clay balls together, that results in '1 sticky' contact. Bringing 3 sticky balls together in a triangular pattern causes 3 'sticks', 3 times as many as the previously 1 stick case. And if 4 sticky balls are brought together in a tetrahedral pattern, that causes 6 'sticking points', twice as many as the previously 3. But **not four times** as many 'sticks!'

So the quadrupling of energy output in the 4-nucleon fusion case, compared to 3 nucleons, is a surprise at first glance. But on closer examination, we note that when 4 spheres touch in a tetrahedral array, that results in the formation of **4 planes**, each consisting of 3 nucleons; unlike just a single triangular array of 3 nucleons forming only 1 plane. So we note the basic geometric shapes and structures involved, and thus the implication that those shapes may relate to the four-fold fusion energy jump and to our modeling.

1.2. (Optional) Comment on Others' Helpful Work

There exist good but rather complicated models and rules for estimating fusion mass losses as more and more nucleons are added to a nucleus to increase its mass, and depending on whether they are protons or neutrons, and other factors. And which proton and neutron combinations will still result in a stable nucleus. Model complexities are not surprising. For example, two protons and a neutron make the rarely occurring Helium-3 isotope that is stable. But two neutrons and a proton make a rarely occurring Hydrogen-3 that is not quite stable. And there doesn't seem to be any stable nuclei with 5 or 8 nucleons.

Despite the challenging task, some scientists have devised rather good and effective models in the 20th Century and early into the next, although models generally involve more than a few rules and are somewhat complicated [2-6}.

The author's supplemental model is very crude and limited in focus. But its focus is especially important. It does not even include charge or spin as considerations. But hopefully, readers will find it is interesting and helpfully thought provoking.

Some of Author's previous papers have shown correlations arising when big spheres surround small spheres in patterns. Volume ratios arose and were noted, and they nearly matched major particle mass ratios in nature. And a slight hint arose that crevice size between three or more equal touching spheres may very loosely correlate with small mass differences among slightly different particles in the same particle class [7-8].

2. Description

2.1. Interpreting Fig. 1

We'll start by discussing Fig. 1, because an aspect of it reveals a surprisingly close equality between a geometric numerical outcome and an empirical outcome as referred to earlier in our Abstract. In Fig. 1 we note the donut hole between the 3 touching nuclei or spheres. That geometry relates closely to e,p-binding energy.

Important Geometric Ratio: Vol. of 2 small spheres each barely fitting through the 'donut hole' formed between 3 bigger spheres --- to the total Vol. of those 3 surrounding big spheres: 0.00246822/ 1

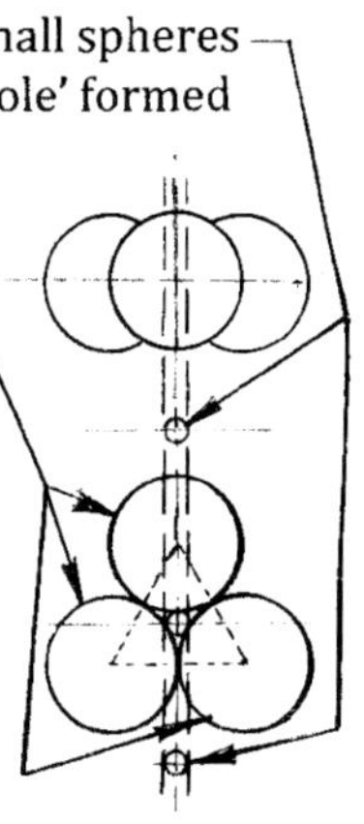

Compare that Ratio with one obtained using the following interpretation: (one among a few rather similar ones) The ratio of the 'e,p-based binding energy' of the atom, Helium-3, after nucleon fusion of 3 Hydrogen-1 atoms -- to the total mc² energy of the Helium-3 atom: 0.00246874/ 1

That 0.00246874/1 ratio is very close to the geometric ratio, 0.00246822/1 with an error of about 1 part in 5000.

Fig. 1; Three Spheres or Nucleons – in a '1 triangular plane', having about ¼ th the binding energy of four nucleons in a tetrahedral array with 4 triangular planes.

But before further detailing our binding energy treatment, we unfortunately need to discuss the presently accepted 'atomic mass' standards, and how to use them. Although awkward, the present standards give sufficiently precise published tables of relative masses. And scientists must generally use them to helpfully calculate the mass difference between empirically measured initial masses and the final fused mass, in other words, the 'mass lost', and 'energy emitted'. So we'll also use all that here, and illustrate its use.

Roughly speaking, since the mid 1960s, the standard has been based on first assigning an 'atomic mass' amount to the 12 fused nucleons of carbon plus their 6 'orbiting electrons' (the common atomic carbon isotope ${}_6C^{12}$). That is assigned 12.000000 u, i.e., 12 units of mass. (Before 1961, the standard was based on the ${}_8O^{16}$ oxygen isotope and we would refer to that atom's 16.000000 'atomic mass units' ('amu').

The present ${}_6C^{12}$ standard's 'whole number', 12.000000 u, is now especially nice for carbon, but still results in a somewhat awkward relative mass value for the common hydrogen atom, a proton with orbiting election, namely 1.007825032 u, instead of a 'tidier' 1.000000 u. And the mass of the proton alone, therefore, comes out to almost exactly 1.007276467 u. Fortunately, relative numerical proportions are maintained, despite the shifts, so we can continue to use the table effectively and accurately enough.

Let us now proceed to calculate the e,p-based binding energy for the one ${}_2He^3$ (Helium-3) atom, made from three ${}_1H^1$ atoms, and as roughly shown in Fig. 1. (Note, one nice aspect of our methodology is this: We start with a simple system of 3 neutral simplest atoms and end up with a neutral system of one simple atom -- and we don't need to speculate about how deeply one formerly orbiting electron got jammed into the nucleus. Or even if it really formed one true, complete 'neutron' in there.)

The initial ingredients, three hydrogen atoms, equals a total mass of 3.0234751 u. The final fused product, one $_2He^3$ atom, has a mass of 3.0160293 u. The mass difference, and thus mass 'lost', is 0.0074458 u. And that is considered the e,p-based binding energy, i.e., the mass equivalent of energy emitted.

(Optionally, one can divide that 0.0074458 u by the mass of the final fused result, 3.0160293 u, to obtain the .00246874/ 1 ratio shown in Fig. 1.)

Our main model for the above reality includes a slight subtlety, but it only affects the outcome very slightly. We imagine 3 abstract balls of energy in space, perhaps spherical fields or ethereal balls, as equal to the mass equivalent energy of that Helium-3 atom, 3.0160293 u. (We can divide that by 3 to get the average mass equivalent energy of each of those abstract balls.) Twice the mass equivalent energy of a single small ball that would barely fit through that surrounding 3-ball array's donut hole is 0.0074442 u. That geometrical result is about 0.021% less than the 0.0074458 u value of our e,p-based binding energy for the event -- thus an error of only about 1 part in 5000.

(Optionally, one can divide that geometrical result, 0.0074442 u, by the total mass of the final atom, 3.0160293 u, to obtain the other ratio, 0.00246822/ 1, shown in Fig. 1, for comparison.)

Suppose for the final mass of the 3 fused spheres above, we had imagined the value reduced by 2 electron masses to make 2 remotely orbiting electrons. That would have only slightly reduced the size of the donut hole in the slightly reduced size of the 3 surrounding big spheres, thus increasing our error to about 0.06% instead of about 0.02%. That's still a very small error.

Suppose we had asked, instead, "What fits through the initial donut hole in the 3-nucleon array before the array fused, instead of what fits through the donut hole after fusion or an abstract after-fusion representation?" Then we could consider, say, each of the three initial nucleons to be the average of 3 positive protons, but with the mass of one negative electron also added to the mix. Twice the mass of one small sphere that would barely fit through that initial 3-nucleon array's donut hole -- equals 0.00746076 u. That comes out 0.2% higher than the e,p-based binding energy, ref. 0.0074458 u, but that is still very close!

A word of caution about the remarkable closeness achieved, especially for the case of the 'after fusion model' calculations: Of course, the closeness could be 'just a coincidence'. Based on what appears to be the amount of darkened volume in the tunnel region, we might expect that that dark volume might be associated only very roughly with such a mass loss. But suppose the numerical correspondence above is not just a coincidence: Then the donut hole dimensions, its unique 'quantum size', and the somewhat strange quantum propensities in our micro-world -- may actually act to generate quantum-sized increment results in our above-discussed case!

Note this for the first model discussed above (the most abstract model treated somewhat subtly): We extended the incompressibility and uniform density concepts in the Bohr 'Liquid Drop Model' to also apply to our ethereal energy spheres or spherical fields. Our treatments also accepted the rather common theme in physics that sometimes a particle or its energy behaves as if it is not in a simple single place. Examples of that are 'electron double slit diffraction' and in chemical 'resonance bonds'. And in systems with potential energy or where there arises more than one degree of freedom or expression of motion.

2.2. A Better Binding Energy Concept and Fig. 1

Let us now discuss, in more detail, what we have termed, our new e,p-based Binding Energy method, and some of its advantages over the old n,p-based Binding Energy method:

First, we will note the problems that arise when the old **n**,p-based binding energy treatment is used to analyze two slightly different 3-nucleon atoms. The first is the stable Helium-3 atom with ($_2He^3$) nucleus, which the old treatment says consists of two protons and one neutron. And the second is the decay-prone Hydrogen-3 atom with ($_1H^3$) nucleus, which the old treatment says consists of one proton and two neutrons. The old binding energy treatment starts with the following basic ingredients: To make ($_2He^3$), 2 hydrogen atoms and 1 unstable neutron are envisioned as used. And to make ($_1H^3$), 1 hydrogen atom and 2 unstable neutrons are envisioned as used. The binding energy is based on the mass difference between the starting unfused ingredients and the final fused atom's mass.

Thus the old binding energy calculations, using empirical data, result in 'binding energies' for those atoms' nuclei, as follows:

For Helium-3, ($_2He^3$): 0.00**82**857 u.

For Hydrogen-3, ($_1H^3$): 0.00**91**056 u.

Note the substantially **greater** '**binding** energy' value for the unstable Hydrogen-3 nucleus compared to the stable Helium-3 nucleus that is thus sadly obtained above.

That is a **disaster** for the commonly used old binding energy method, since the Hydrogen-3 nucleus is unstable and decays (half life about 12 years), but the Helium-3 nucleus is stable. Despite the old method's predictive term, 'greater binding energy' (which implies a greater binding strength should result), the reality is that a significantly weaker bond results for the decay prone ($_1H^3$).

Somewhat relatedly, readers are reminded that when, say, hot liquid tungsten starts cooling and crystalizes, it gives off much greater 'heat of fusion' than other crystalizing metals. And we customarily and correctly associate that much greater heat of fusion with the much higher tensile strength of ordinary tungsten, compared to other metals.

So the empirical instability of ($_1H^3$), despite the old binding energy method assigning it much greater binding energy than stable ($_2He^3$), is contrary to the old method's predictive language and its associated relative numbers. And the method is contrary to practices and outcomes when treating other analogous physics topics, such as heats of fusion. So all that is **not** acceptable. It at least indicates a great need for an alternate binding energy methodology to be made available, an alternate way of handling and interpreting empirical data. (Of course, that's not to say the old method is never helpful – an example of that is given later.)

Now let us see how our advocated e,p-based binding energy method treats those atoms, and obtains better predictive results: Our method's non-fused starting ingredients for making both Helium-3 and Hydrogen-3 are the same, a neutral system of (3) Hydrogen-1 atoms. The difference in mass between that and masses of the final fused atoms are our binding energies, and are as follows:

For Helium-3 nucleus, ($_2He^3$): 0.0074<u>4</u>58 u.

For Hydrogen-3 nucleus, ($_1H^3$): 0.0074<u>2</u>58 u.

Note our e,p-based binding energy treatment gives a slightly greater binding energy for ($_2He^3$) than for ($_1H^3$). That is very appropriate and good. The implications of the new now greater <u>binding</u> energy for ($_2He^3$) compared to ($_1H^3$) implies literally that ($_2He^3$) is '<u>bound</u>' together more tightly and is apt to be more stable than the ($_1H^3$). And it is, since the ($_1H^3$) is unstable!

But the fact that the new binding energy of ($_2He^3$) is still not huge nor a lot greater than for ($_1H^3$) means this: That the nucleus of the more common helium isotope ($_2He^4$), that we associate with the tetrahedron array, is likely to be a lot more common than the ($_2He^3$) that we associate with the triangular array. And indeed, that much greater occurrence of Helium-4, is the reality in our solar system and likely for the world.

Our e,p-based binding energy method is, in effect, like making the hydrogen atom a major mass standard for helping to calculate and to judge relative 'mass loss' when various fusions occur. That standard might not be perfect nor beyond improvement, but historically the Bohr Atomic model for the hydrogen atom provided a great basic leap forward. It is understandable using 8th or 9th grade algebra. And no other atom came even close to being so easily treated and still revealing so much. And as said previously, the hydrogen (proton) nucleus is by far the most common nucleus in the universe and does not, itself, incorporate a neutron nor need to.

It is the stable embodiment that serves vastly more often than others as the basic positive charge. And similarly, but oppositely, the electron serves so often as the negative charge.

So why not compare the mass of, say, two simple hydrogen atoms (net charge zero) with the resulting mass of a 'deuterium' atom (net charge also zero)? And then asking the basic question, "How much conventional mass was lost by that 'transition' or replacement?" That is a very self-justified question or analytic tool in its own right, anyway! And similarly for treating atoms with more nucleons by just imagining more Hydrogen-1 atoms used at the start.

It is a great 'systems-before' and 'systems-after' analytic tool, without scientists being 'side-tracked' by the overly compulsive immediate need to know (or think they know) exactly where the 'bound' electron is. We don't need to know exactly where the electron is that gives each system its net electrical neutrality.

Nor do we even have to say, 'the neutron and proton inside the nucleus' vs. 'the neutron and proton outside the nucleus', since the masses in each case are different, anyway, depending upon whether each is 'free' or fused into a bundle. And using similar names for different things (and different masses) tends to lead to unnecessary verbal, conceptual, and philosophical contradictions, too. The problem is due to the overly compulsive propensity to rush in too fast and <u>literally</u> 'over <u>micro</u>-manage' the (micro) system. Or what one thinks is the system.

Consider our preferred alternate method instead: If the electron is orbiting the nucleus, fine. If it's touching the nucleus, fine. If it's somewhere inside the nucleus, fine. Or if at some time it's, so-to-speak, jammed far into the nucleus, that's fine! (All our e,p-based binding energy method needs to know is that the 'beginning system' is electrically net neutral and that the 'final system' is also electrically net neutral.)

Ultimately, the net fusion of several simple Bohr hydrogen atoms to form fused geometric formations, like Helium-4, and even more massive bundles - may increase slightly the 'ether' in our universe. And the reverse action may slightly decrease the amount of ether. Perhaps that is somewhat analogous to the earth's ecosystem: Oceans partly evaporate to give us high humidity; then rain, and that back into the rivers, lakes, and oceans again. (And, hopefully, the greater temperature stability that brings us.)

Our e,p-based binding energy method often gives us a lot of useful information without the problems that often arise when scientists attribute or impose on nature -- their own, perhaps, narrow minded viewpoint or overstretched attributions. Hopefully, we'll <u>first</u> view nature by allowing it to 'do its own thing its own way' -- not necessarily our mentally conceived-of way, and see what we can first glean from that.

We previously applied our new e,p-based binding energy method to atoms with <u>3</u> fused nucleons. But when Sears and Zemansky first introduced their e,p-based binding energy, they applied it to <u>4</u> fused nucleons, the Helium-4 atom [9]. They wisely imagined for the mass of their starting ingredients, a neutral system of 4 unfused Hydrogen-1 atoms; and for their neutral end product, 1 Helium-4 atom. Then they calculated the mass difference. That gave them the energy emitted, the binding energy.

2.3 Interpreting Fig. 2

We now discuss in more detail -- the common 4-fused nucleons case, often designed 'Helium-4'. See Fig. 2 below:

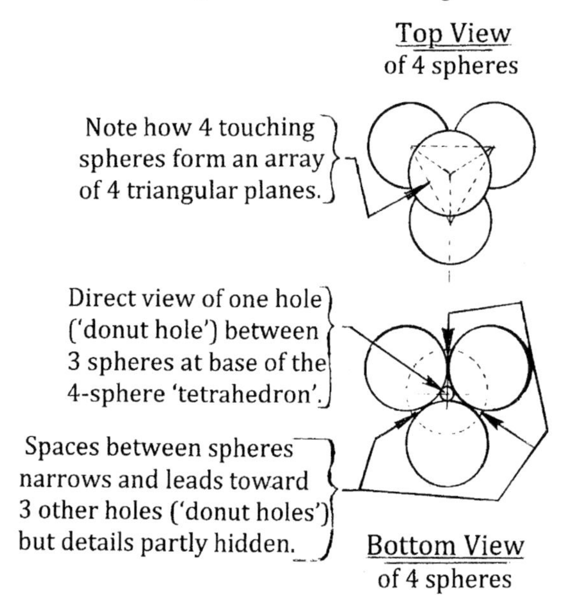

Fig. 2; Four Spheres or Nucleons – in a tetrahedral array of <u>4 triangular planes</u>. Has about <u>4 times</u> the binding energy of 3 fused nucleons in a <u>1 triangular plane</u>.

As previously noted, we can geometrically imagine Helium-4 as resulting from adding a 4th nucleon on top of a 3-nucleon triangular array, thus forming a tetrahedron shaped 4-nucleon array. And as previously noted, we can thus imagine it to consist of 4 triangular planes, each with a donut hole, totaling 4 donut holes. That is to be contrasted with only 3 nucleons, as shown in Fig. 1, forming only 1 triangular plane or array, with only a total of 1 donut hole.

We have previously advocated the logic of likely associating that increase, from 1 donut hole to 4 donut holes, with a roughly 4-fold increase in fusion energy emitted, or fusion mass lost. Empirically, the increase in mass loss comes out a little less than the simple 4-fold jump we envisioned.

Using the standard old n,p-based binding energy method, the increase is interpreted to be from 0.0082857 u to 0.030378 u, about 8.34% less than a full 4-fold jump. But using our e,p-based binding energy method; the increase is interpreted to be from 0.0074458 u to 0.028698 u, and thus only about **3.64% less** than our 4-fold increase envisioned!

We used the wording, 'our 4-fold increase envisioned'. But actually, the somewhat complicated interior cavern (inside our tetrahedron shaped 4 nucleon array) does not really allow us to be as confident about our roughly 4-fold prediction as we might wish. On the one hand, there are aspects about that interior and possible flows, there, that might lead one to predict a somewhat less 'tunnel effect', because of turbulence developing. And since there are 'close quarters' in the interior (instead of the nearly 'open skies' on the exterior), that might also impede the effect.

On the other hand, the internal tunnels seem likely longer, in a sense, than for the simple 3-nucleon case in Fig. 1. And that might increase the tunnel effect. Exact attempted calculations of interior tunnel effects are not given in this paper, but the author is actually somewhat pleasantly surprised that the net empirical result comes out as close to being a '4-fold' increase as it does!

There may be a little 'balancing-out good luck' involved, and perhaps the rather good net result implies that a 'quantum action', by nature, is still somewhat present. But not so overridingly powerful as to dwarf the effects of other factors.

2.4 Interpreting Fig. 3

This author does not here present a detailed, accurate drawing for modeling the nuclear bond between the two nucleons comprising a deuteron. So Fig. 3, below, is just the barest roughest outline of, say, two fused nucleons:

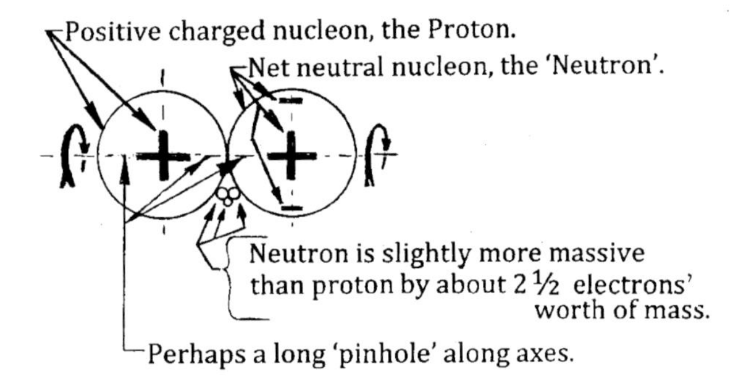

Fig. 3; A Deuteron, two Nucleons, or two Spheres. Any other sketch details totally conjecture, not really known.

Historically, however, quite a few contributors to the 'Proceedings of the Natural Philosophy Alliance' and the journal, 'Galilean Electrodynamics', have presented respectable and thoughtful models. But details of those models are beyond the scope of this paper. Of course, one challenge has been to show and describe the details of what is really going on, that is the cause of such particle behavior -- which scientists have had to concoct the 'charge algorism' to treat efficiently.

Strangely enough, the e,p-based binding energy associated with the relatively weak deuteron's bond does not seem magnitudes greater than energies associated with the electron particle. Nor much greater than the energy associated with point charges separated by a nuclear distance (10^{-15}m). (Even though the neutron and deuteron are considered to have no or trivial dipole moments!)

Interestingly, the energy emitted when just 2 nucleons fuse to form a deuteron is about 2.82 times the (mc^2) energy of one electron mass. (To express that in terms of our e,p-based binding energy, we say this: The binding energy of the deuteron is about 0.00154829 u compared to about 0.00054858 u for a 'free electron'. That comes out to about 2.822 times the energy of the electron.)

But consider that compared to the following related energies: The energy associated with the radius of 'classical electron' (radius = 2.82×10^{-15} m) vs. the energy associated with bringing a 'point' electron 'from infinity' to barely fully inside the surface of a proton (say 1.0×10^{-15} m distance from proton's center). That latter energy is also about 2.82 times the (mc^2) of one electron mass. The likely radius of a proton is about 1.2×10^{-15} m. Those are approximate estimates.

All that is merely an argument for the likelihood that the relatively weak binding between the two nucleons of a deuteron is mostly just 'electric flow' related. That is, related to flows associated with just one or a very few basic units of charge.

Some scientists have also noted the following: Suppose we bring the electron from a 'classical distance' of 2.82×10^{-15} m away from proton -- to the proton's surface, (proton's radius is about 1.2×10^{-15} m). That would constitute adding enough mass equivalent energy to the system to equal the slight difference in mass between a neutron and proton. That energy difference is not much less than the deuteron's e,p-based binding energy.

2.5. More insight into Binding Energy using Deuterons

The deuteron may offer a very special insight into the pros and cons of our simple e,p-based binding energy method and the old n,p-based binding energy method. That is because real experiments have been done to see how much energy is needed to break the deuteron into one neutron and a proton, using a gamma ray.

At first glance, the result and conclusion seem very quick and simple -- too simple. A strong gamma ray, with 'mass equivalent energy' of about 4-1/2 electron masses does the job. Say, we hit 100 deuterons with 100 of those gamma rays. That breaks up each; that 'does the trick' in less than a fraction of a second. So, some scientists might say, "End of test, turn out the lights, go home, and publish the conclusion: The deuteron has 4-1/2 electrons worth of binding energy, a pretty strong bond!" The old binding energy method and its approach seem great, at first glance.

But suppose more diligent experimenters were willing to stay and observe the results for more than 12 minutes, the mean half-life of neutron. They would further note this: After the first deuteron was broken up by the very strong gamma ray, into the proton and neutron, the neutron, itself, broke up into a more basic proton and electron. And gave off about 1-1/2 electron masses' worth of energy back to the experimenter. Imagine the experimenter using that feedback to 'boost' the energy of, say, a next much weaker gamma ray. Say, that next gamma ray, with only 3-electron masses worth of energy initially - gets boosted back up to the 4-1/2 electron masses worth of energy, to split the next deuteron!

And so on, say, for the 99 other deuterons, i.e., saving 1-1/2 electron masses worth of energy per each of those splits. The beauty of the simple e,p-based binding energy method is this: In effect, it gives a weaker net gamma ray energy needed, only 3 electron mass equivalents net worth of gamma ray energy needed for each subsequent deuteron breakup, not 4-1/2 electrons.

So that non-wasted energy or cogeneration 'feedback' result also shows how to do more with less energy. And, as said, that lesser energy need is clearly indicated by our e,p-based binding energy calculation for the deuteron, instead of being obscured by the old n,p-based binding energy calculation.

In general, naturally occurring radioactive elements do not emit neutrons - except in very rare cases of 'spontaneous fission' or when hit by an incident particle. Various naturally occurring radioactive elements typically transform by emitting an electron or by emitting an alpha particle, the nucleus of Helium-4.

2.6. Comments on Binding Energies of more than 4 Nucleons

As the number of nucleons in a bundle increases, a variety of complex subtleties seem to manifest themselves more strongly. So the reader is best referred to the references at the end of this paper for treatment of those subtleties. Still, we will discuss some hopefully helpful analogies, and make some other general comments -- about bigger and bigger nuclei, as follows:

We earlier noted the three major basic ways that nucleons can fuse to form a nucleus: #1) The weak low energy bond between two nucleons like the deuteron in Fig. 3, which we'll term 'electric flow related'. #2) The medium energy or medium strength bond between three nucleons like in Helium-3 in Fig. 1, which is likely related to its having only one donut hole. #3) The high energy or high strength bond like in Helium-4 in Fig. 2, which is likely related to its having four donut holes.

But as we add more nucleons to most other nuclei, we note the following about the jump in bonding strength or energy per nucleon added: The jump is usually a 'hybrid' or mixture of those bonds in the previous paragraph, for example, somewhere between a low and medium energy jump. Or as still more nucleons are added, between a medium and high-energy jump.

That has an analogy in the 'world of chemical bonds' -- where such descriptions arise as 'partial ionic character of a covalent bond'. Or for covalent benzene bonds, a compromise carbon-to-carbon bond length 'with 50% double-bond character'. And so many important chemical bonds exhibit the related so-called 'resonance behavior', somewhat like benzene.

Roughly speaking, in the range of 5 to 8 total nucleons - the binding energy jumps are erratic, averaging in the 'weak electric flow class', but occasionally rising to an almost 'one donut hole' class. From 9 to 12 nucleons, jumps are somewhat beyond the one donut hole class - as if to include a slight 'four-donut hole character' in the mix. From 13 to 16 nucleons - we note slightly smaller binding energy jumps than in the previous range.

Generally after that, as nucleons are further added until about 60 nucleons are accumulated; the 'binding energy jumps' increase, as if to indicate a slightly greater four-donut holes character. But after about 60 nucleons, the magnitude of the jumps becomes less, most likely because of the ever-growing repulsions of the increasing number of positively charged protons in the nucleus.

The lack of any stable 5-nucleon nucleus may partially relate to the following: Plopping the 5th nucleon on top of any one of the four 3-nucleon triangular arrays (comprising Helium-4) - might stop or impede flow involving a pair of the helium's 4 donut holes. And block the former 'port to port' line-of-sight that was partially present. The added flow between a resulting new pair of donut holes would be, at best, very asymmetric relative to previous flows, and in close proximity to them too. Some of that problem might exist even if a 5th nucleon is added anywhere onto the 4-nucleon bundle.

Those problems seem to be much less, for the case of an 8-nucleon bundle, and that seems to have enough appeal. And the 8th nucleon does, in fact, seem to stick on the bundle for nearly a second. But it seems likely that if it and another existing nucleon slide even slightly away from their ideal position, the beginnings of two tetrahedral formations will appear. And perhaps the appeal of two independent 4-nucleon bundles competes with the less competitive single 8-nucleon bundle. So it splits up into two 4-nucleon bundles, i.e., two 'Helium-4' nuclei.

3. Conclusion

In about 1960, Sears and Zemanski introduced an alternate and simpler way of calculating the mass loss associated with nucleon fusion. And thus the energy emitted, or the binding energy of a nucleus. Their illustration started with the net neutral system of (4) Bohr hydrogen atoms, and it finished with those fused to form a net neutral Helium-4 atom. They calculated the mass difference between the former and latter, and thus the large energy emitted. That was, in effect, a, useful, alternative 'binding energy' result. That was because they used electrons and protons as their starting building blocks, instead of the old method that used 2 **unstable** neutrons and 2 hydrogen atoms as the starting building blocks.

We called their new method the 'e,p-based binding energy' method because it uses the **e**lectron and **p**roton (or the Bohr hydrogen atom) as its building blocks, instead of the old n,p-based binding energy method's use of the neutron and proton with orbiting electron.

The new method may be used for calculating the binding energy of the net neutral ${}_2He^3$ and ${}_1H^3$ atoms by starting with the net neutral mass of (3) ${}_1H^1$ Bohr atoms. And the excellent results show a slightly greater e,p-based binding energy for the stable ${}_2He^3$ nucleus than for the unstable ${}_1H^3$ nucleus. That contrasts with the old system's confusing and dubious result, which calculates a much greater binding energy (or binding strength) for the **unstable** $_1H^3$ nucleus!

Most of this paper's main themes are, with some effort, even evident using the old binding energy method. But the themes are more readily evident using the newer e,p-based binding energy method.

We have shown that Helium-4 has about 4 times the nuclear binding energy compared to Helium-3. We theorized that that is because a tetrahedral shaped Helium-4 nucleus has 4 times as many triangular planes, each with 1 hole, like in a donut -- compared to Helium-3 with just 1 triangular plane with 1 'donut hole'.

We also showed that the fusion energy emitted after (3) Hydrogen-1 atoms fuse to form a Helium-3 atom is as follows: Twice the energy of one small ball that could barely fit through the donut hole in the array of 3 bigger touching balls. That is, assuming each of those 3 bigger balls has one-third of the energy of the final fused Helium-3 atom. That geometry-based estimate vs. the atom's e,p-based binding energy -- is accurate to a small error of about 1 part in 5000.

There are also other methods in the paper for treating the above fusion and they are more detailed and less abstract. Their predictions give nearly the same result as the above with only slightly greater error. All of our methods use the approximations inherent in the Bohr 'liquid drop model' of the nucleus. That is virtual incompressibility and uniform density, regardless of whether applied to nucleon spheres or 'ethereal spheres of energy'.

It is, of course, quite possible that nucleons are not precisely 'billiard ball' shaped, nor are 3 or 4 of them so arranged as to form perfect triangular or tetrahedral patterns. But perhaps the ethereal fields around them or near them do nearly achieve those ideal forms! More comment on that also given in the paper.

It is hoped that eventually e,p-based binding energies will be calculated for almost all nuclei; that various graphs will be made based on that; and that a lot of nuclear science will be reexamined based on those results. But that is beyond this paper.

It also might be interesting to try to explain the following in simplest terms: Some modern listings give the masses for the free proton, free electron and ground state of the Bohr atom in 'u' units [10]. But that data seems to indicate that the mass of the free proton plus the free electron equals very slightly more than the mass of a Bohr atom. That seems to imply, for some systems, that a motion-related 'relativistic' mass decrease occurs for the electron, instead of an expected increase. But perhaps the data is not quite as precise as hoped, or some alternate explanation.

References

[1] http://en.wikipedia.org/wiki/Francis_Sears
http://en.wikipedia.org/wiki/Mark_Zemansky

[2] Lucas, J., "A Physical Model for Atoms and Nuclei," *Galilean Electrodynamics*, vol. 7, no. 1, pp. 3-12 (Jan./Feb. 1996).

[3] Mayer, M. G., and Jensen, J. H. D., *Elementary Theory of Nuclear Shell Structure*, John Wiley & Sons, New York (1955)

[4] http://en.wikipedia.org/wiki/Semi-empirical_mass_formula
(The Weizsacker's formula or the Bethe-Weizacker formula)

[5] Pauling, L., *General Chemistry*, Nuclear Chemistry, 26-7 to 26-10, Dover Publications, Inc., Mineola, New York (1988).

[6] Borg, X., http://www.blazelabs.com Home, Particle Part 2, Magic Numbers Explained

[7] Littmann, C., "Particle Mass Ratios and Similar Volume Ratios in Geometry", *J. Chem. Inf. Comput. Sci.* 35 (3) 579-580 (1995)

[8] Littmann, C., "Volume Ratios in Patterns vs. Mass Ratios of Prominent Hyperons and Some Other Particles", *Proceedings of the NPA* **10**, 166-169 (2013)

[9] Sears, F., Zemansky, M., *College Physics*, Radioactivity and Nuclear Physics, 49-10, 3rd Edit., Addision-Wesley Publishing Company, Inc., Reading MA (1960)

[10] http://en.wikipedia and enter topics, 'Electron', 'Proton', 'Isotopes of Hydrogen', in search box provided. Note, data provided may change since this paper was drafted in Dec., 2014.

Hubble's Law and Dark Energy are False

Al McDowell
4425 Touchstone Forest Rd, Raleigh, NC 27612
e-mail: almcd999@earthlink.net

The universe is commonly believed to be expanding due to Dark Energy from a Big Bang in an unique manner prescribed by Hubble's Law. Astronomy data and the Cosmic Microwave Background Radiation experiments contradict all of these beliefs. Dark Energy does not exist, Hubble's Law is false, the Big Bang is preposterous, and the universe is not expanding.

1. Introduction

Hubble's Law states that the galaxies of the universe are moving away from Earth at velocities in direct proportion to their distances from Earth. The physics profession further believes that a mysterious force called Dark Energy is causing the expansion. Their empirical evidence for these beliefs is based on galactic radiation redshift, which they assume has lower frequencies and greater wavelengths in direct proportion to galactic distances from Earth. This redshift explanation would be valid if all redshift were due to the Doppler velocity of galaxies moving away from Earth.

However, it is shown below that all of these common beliefs are totally false.

2. Astronomy Data Contradicting the Hubble Law

Red light has a lower frequency than blue light. Redshift (or blueshift) is the observation of light at lower (or higher) frequencies than we would see in the lab when emitted by the same type of atoms not moving relative to the observer. The redshift of galactic light is the basis for the mainstream structure of the universe.

The well-known astronomer, Halton Arp, has discovered and photographed numerous high-redshift quasars attached to nearby low-redshift galaxies by thin filaments of extremely fast material that normally appear to be leaving a pole of the galaxy center. This information can be found in his book, *Seeing Red: Redshifts, Cosmology and Academic Science* [1] and on his website *www.haltonarp.com*. Each galaxy and its connected quasar are obviously almost exactly the same distance from Earth, and both must be moving at nearly the same velocity through space.

On the contrary, Hubble's Law states that all redshift is due to Doppler velocities and that the distances to the light sources are directly proportional to their velocities moving away from us. This means that a galaxy and its connected quasar would have to be not only moving away at the same velocity, but would also have to be the same distance from Earth, even though they have the same redshift, which is impossible.

Hubble's Law defines a very specific manner of universe expansion in which no center can be identified. This type of expansion is commonly described as like a loaf of rising raisin bread, in which the distance between any two raisins is directly proportional to their velocity moving away from each other. If there had been a Big Bang, and if the expansion did not obey Hubble's Law, then we would observe an apparent center of the universe on which galaxies would be moving away from us faster on the other side of the origin. No such evidence of a center has been found.

In the conventional structure of the universe, distances to galaxies are extrapolated from Hubble's Law and the redshift measurements. This Law implies that the furthest galaxies we see are moving away from us at either nearly the speed of light or several times the speed of light, depending on whether you assume that Special Relativity is valid.

However, the data from Arp show that the Doppler velocities of distant light sources moving away from Earth cannot possibly explain all of the cause of redshift of distant light. There has to be at least one additional cause. We have no way to know how much actual observed redshift is due to velocity and how much is due to something else.

Our most accurate estimates of cosmic distances are for a particular type of star called Cepheid Variables, whose luminosities pulse over periods of several days. From the pulsation period, we can estimate the luminosity of the star and estimate its distance from the comparative brightness we see on Earth. Arp finds that the redshift of the Cepheids correlates with distance over relatively short ranges, but their redshifts become too large to explain by the Hubble Law at longer distances, especially for "high luminosity Class I galaxies." Like quasars, Cepheids at longer distances seem to be a class of star that has an intrinsic cause of redshift in addition to that of galaxies.

A UC, Santa Cruz study of 14 gamma ray bursts (GRBs) found that they all appear to have galaxies directly between them and us [2]. This unlikely coincidence can be explained by ignoring the Hubble Law and recognizing the obvious likelihood that the GRBs originated from within the galaxies with redshifts larger than the rest of the galaxies.

Stars often appear in pairs, chasing each other around their orbits. For many binary stars, particularly Wolf-Rayet binary stars, the average redshifts of one star averaged over the two sides of their orbit going away and toward us are substantially different than the average redshift of the other binary star, implying that one of them is moving away from us faster than the other, an impossibility for binary stars locked in orbit around each other. This further contradicts Hubble's Law.

The quasar, Cepheid star, GRB, and Wolf-Rayet binary star evidence seems to confirm that velocity is not the only cause of cosmic redshift. This compromises the ability of Hubble's Law to predict velocity or distance based only on redshift.

3. Galactic Redshift Periodicities

Astronomers W.G. Tifft, A. S. Szalay, T. J Broadhurst, B.N.G. Guthrie, W.M. Napier, E.M. and G.R. Burbidge [3, 4, 5, 6, 7, 8, 9] and others have found periodicities in the redshift from galaxies. This data is from both 21 cm radio telescopes and from optical telescopes. Applying Hubble's Law to these data implies that these galaxies are organized into concentric shells with unique velocities and distances for each shell, with Earth at or near the center of the universe. This Earth-centered concentric shell model of the universe seems absurd.

A more likely implication would be that these concentric shells exist only in local areas within the universe as remnants of some form of local explosive event. However, even this implication would require the validity of Hubble's Law.

Electromagnetic radiation obtains its redshift either from the source that emitted it or from the velocity of observers in the direction of the source. It cannot obtain redshift in transit. Consider dropping marbles at the frequency of one per second into a tilted 6-ft long plastic tube. When the tilt is increased, the marbles roll faster and their spacing increases, but the frequency of their arrival at the low end of the tube remains precisely the frequency at which they were inserted.

This means that at least some of the galactic light must be quantized into periodic frequencies at the source. This could be periodic noise in the electron orbits in the atoms that emit the light, but this is not yet known. Whatever the source, the periodicities cannot represent concentric shells of galaxies without the validity of Hubble's Law.

4. Cosmic Microwave Background Radiation

Most physicists believe that the Cosmic Microwave Background Radiation (CMBR) comes from the "first light," which presumably appeared 400,000 years after the Big Bang 14 billion years ago. To be reaching Earth now, the matter had to travel at about 35,000 times the speed of light to get 14 billion light-years from the Big Bang before it braked suddenly to below light speed and radiated the first light. This is NOT believable!

If there had been no Big Bang, there would have been no first light, and the CMBR would come from matter such as hydrogen distributed throughout the universe. If the Big Bang had occurred, the first light and the CMBR would have originated throughout the universe, not only at the outermost distance. Moreover, the CMBR is weak thermal radiation from atoms at 2.7° K. The brilliant starlight from galaxies must surely travel further than the CMBR before becoming too dim to see. Thus, the CMBR must come from sources distributed throughout the universe rather than from the distant edge of the universe.

If the universe were expanding as Hubble's Law says, when our telescopes look in any single direction in the sky, the light from distant atoms would be redshifted more than the light from nearby atoms. However, the distributions of CMBR wavelength intensity follow the Planck blackbody thermal radiation distribution perfectly, meaning that there is no redshift in any given direction. The lack of CMBR redshift from all distances in a single direction implies that the universe is neither expanding nor contracting, contradicting Hubble's Law.

5. Conclusions

The failure of Doppler velocities to explain the redshift of galaxies begs the question of what does cause the obvously very real redshifts of galaxies. There are broadly two choices. Either galactic radiation loses frequency as it travels, or galactic light leaves galaxies with "intrinsic" redshifts established by the differing enviroments of the galaxies that cause the emitting electrons to be at different energy levels in different galaxies, almost certainly based on their age. This author prefers the latter alternative, and he discusses the several alternatives in his book *Uncommon Knowledge: New Science of Gravity, Light, the Origin of Life, and the Mind of Man* [10].

Regardless of the real cause of redshift, the evidence presented here enables the logical conclusions that:

The Big Bang is fiction.

The universe is not expanding or contracting.

Hubble's Law is false.

Dark Energy does not exist.

We have little idea how far away the distant galaxies are, which means that our maps of the universe are inaccurate. The galaxies with the most redshift are probably the newest galaxies, and the galaxies with the dimmest radiation are the furthest away among those with a given redshift, but we do not know how far away the galaxies are nor their relative distances.

References

[1] Arp, Halton, 1987, "Seeing Red: Redshifts, Cosmology and Academic Science," Montreal, Canada: Apeiron.

[2] Schilling, Govert, 2006, "Do Gamma Ray Bursts Always Line Up with Galaxies?," Science 313 (August 11): 749.

[3] Tifft, W.G., 1976, "Discrete States of Redshift and Galaxy Dynamics," Astrophysical Journal 206 (May 15, pt. 1): 38-56.

[4] _____. 1996, "Global Redshift Periodicitiesand Periodic Structure," The Astrophysical Journal 468 (September 10): 491-518.

[5] _____, 1997, "Redshift Quantization in the Cosmic Background Rest Frame," Journal of Astrophysics and Astronomy 18: 415 33.

[6] Szalay, A.S., T.j Broadhurst, N. Ellman, D.C., and R.S. Ellis, 1993, "Redshift Survey with Multiple Pencil Beams at the Galactic Poles" Proceedings of the National Academy of Science USA 90 (June): 4853-58.

[7] Broadhurst, T.J., R.S. Ellis, D..C. Koo, and A.S. Szalay, 1990, "Large-Scale Distribution of Galaxies at the Galactic Poles," Nature 343 (February 22): 726-28.

[8] Guthrie, B.N.G. and W.M. Napier, 1991, "Evidence for Redshift Periodicity in Nearby Field Galaxies," Royal Astronomical Society, Monthly Notices 253 (December 1): 533-44.

[9] Burbidge, G. and A. Hewitt, 1990, "The Redshift Peak at x = 0.06, Astrophysical Journal 359 (August 20): L33-L36.

[10] McDowell, A., 2014, "Uncommon Knowledge: New Science of Gravity, Light, the Origin of Life, and the Mind of Man," Bloomington, IN: AuthorHouse.

What is Zero Point Energy?

By Dr. John V Milewski with comments from Dr. Niels Winsor
908 Shelton Court, Wilmington, NC 28412, USA
e-mail: jvmilewski@aol.com

Where does it come from and what is it composed of? That is the common topic today in energy talks. They are all saying there is Free Energy or as others call it Zero Point Energy, but nobody is saying what it is and where it is coming from and of what is it composed? For years now many scientists are saying that yes there is an energy out there in fact we live in a "Sea of Energy" (Moray, 2002) but they still do not define it enough to know what it really is, so we can work with it. It is difficult, elusive, hard to detect and capture. The purpose of this article is to bring definition to this energy so we can capture and work with it.

1. Introduction

When the temperature is reduced to absolute zero there is still a very large amount of energy left in matter or space (vacuum). That energy is defined as Zero Point Energy. An important insight is that it is mostly magnetic in nature and extremely high frequency and extremely high velocity and that's one of the reasons we have not detected it. Most all the energy we use now is electrical in nature and most all our tools for working with it are specifically designed to detect electric energy, in all its many forms. Currently we are working mostly with radiation that we generate, that is electromagnetic (EM) in nature, like light, radio, and TV type signals.

This EM energy is a transverse wave with a large electrical component and a small magnetic component traveling at 90 degrees to the electrical component, and they travel at "C" or the speed of light.

2. Features of This New Energy Defined

I call the new energy Superlight (Milewski, 2009). It is mostly magnetic in nature and in its radiation form is magnetoelectric in nature. This magnetoelectric energy was first defined by Tiller in his paper. (Tiller, 1975) The new magnetic type energy is ME radiation and travels at more than 10 billion times the speed of light. These ME waves are magnetic, torsional, and longitudinal in nature. We are not now equipped to easily detect, measure, or capture it. These ME waves are generated in the core of the large galaxies, and jet out of the axis of rotation, in both directions, perpendicular to the plane of rotation of these galaxies.

[Please see Figure 1 below.]

I believe that galaxies are rotational spirals of stars that are being pushed by gravity (Matthew 2002; and Van Flandern, 2002), spiraling in toward the galactic center. As they get closer to the center, their temperature and density increase to values inconceivably high. This I call the "Schauberger effect" of implosion energy (Coats, 1996)

[Please see Figure 2 below.]

3. Creating a New State of Matter

Thus all the matter is continually being compressed and getting denser and denser and hotter and hotter until, it is my contention, that the matter reaches a critical value of temperature and density, then there is a phase change from electrical in nature to magnetic in nature. Instead of electrons in orbit about a nucleus of protons and neutrons, the new matter consists of magnetic monopoles (which are extremely small particles of energy traveling at extremely high velocity) in orbit about an opposite pole, magnetic based nucleus.

This was discussed in much detail by Newman in his book: The Energy Machine of Joseph Newman (1998). In this book, he is discussing figures (on his pages 10 and 11), in which he describes the magnetic particles as north or south pole magnetic particles (monopoles) that he says travel at the speed of light and spin at the speed of light too. He says the north pole particles come out of the north pole and travel to the south pole and the south pole particles come out of the south pole and go to the north pole, Thus creating the magnetic field that magnets are known for.

[Please see Figure 3 below.]

Also, it is my contention that the permanent magnets get their energy from the zero-point or Superlight field of the Sea of Energy that we all live in. That makes magnets Superlight converters, which have the property of converting Superlight to magnetic monopoles.

What is unique about this new state of matter is what also is created in the center of the Galaxies. When the super-speeding magnetic monopoles change energy states and drop down in orbit, they radiate Superlight (Milewski, 2009) or magnetoelectric energy, that comes funneling out of the rotational axis of the galaxy. Thus it is this Superlight energy that is now coming out of about 400 billion Galaxies, as seen in the deep space photo by Hubble's telescope (Fig. 4) to form the Sea of Energy we now call Zero Point Energy. This is a similar process as to how regular light is generated in electrical matter which have electrons dropping down in their orbits and thus radiating the familiar electromagnetic radiation of light.

[Please see Figure 4 below.]

4. Capturing and Using Zero Point Energy

It is hard now, due to the lack of suitable tools, but as we make new tools, it will slowly become possible. The first thing we have to make is a more efficient wire for transportation of this new energy, which will be in the particle or wave structure of the magnetic monopoles. They are so small and so energetic that most every thing is a room temperature conductor for them. I believe that they travel through most everything at "(C^2)" We have learned from Philip S. Callahan that tree roots are good conductors for magnetic monopoles. (Callahan, 1980) Their magnetic conductor is the wood that most of the root is composed of, and their magnetic insulation is the electrically conductive layer that contains the sap under the bark. Thus from Callahan's work we see the conductors for magnetic monopoles and insulators for magnetic monopoles are just the opposite of what we use now as conductors of electric monopoles or electrons, or as we call it, electricity. So normal electrical conductors will be insulators for the flow of magnetic energy. Similar information can be deduced from the internet writeup titled "Magnetic energy to heal the planet." (Naudin, 2014) They used a magnetic conductor like iron wire that is surrounded by an electrical conductor like a copper plating on the outside. This magnetic conductor was used in all their 14 different experimental apparatuses that run only on magnetic energy.

5. Conductive Wire for Magnetic Monopoles

We think that and ideal wire conductor for magnetic energy would be a small diameter tube-like structure that is a flexible metallic or conductive polymer. This tube is then filled with a magnetic conducting fluid. If needed, the tube is then covered with a metallic or electrically conductive coating. For advanced systems, coating by superconductors will be used.

The magnetic fluid is made of a colloidal suspension of magnetite in an organic oil-like fluid. We choose magnetite over other magnetically conductive materials because it is one of the best magnetic conductors that also has zero hysteresis. This will permit it to handle and transport extremely high frequency pulses of magnetic energy without reversal losses. Also, these high frequency magnetic pulses are needed to produce light and run levitation devices.

6. What's Next after Wire Development?

The development of the wire for conductivity of magnetic monopoles will be followed by circuit development. For these new circuits there will be needed a whole new set of magnetic circuit control devices, such as capacitors, inductors, coils, transformers, resistors, switches, amplifiers, high frequency pulse generators and other circuit controls comparable to what is now used in electrical circuits.

To supplement the application of ultra high frequency devices, a high frequency AC magnetic energy battery, that will be held in a magnetic Faraday-Box-like container. It will be base on resonant magnetic tank circuits. The next stage will be solid state devices. For most of this work there is no conventional scientific theory for these ideas. So when you are working to solve a way-out problem we have to think outside the electrical box.

7. Applications and Performance for the Home

This new energy will be generated on-site for each application, therefore there will be no need for central power plants and big distribution systems. The average home will have its own power supply unit of magnetricity, probably able to produce an equivalent of 5 to 10 KW of electrical power.

This unit will be a super-high-frequency magnetic pulse generator. With multi-frequency capability, different frequencies will be used to run different energy systems. For lighting there will be panels that hang on the wall, like picture frames. Home heating units will be built into the walls or floors or hang like drapes or lie on the floor like carpets.

These same units can be used for cooling too. Each appliance will also have its own frequency and power requirements supplied from the one central resonant power unit. The cars will run on magnetricity which will keep them charged whenever they are parked in or near the home. Their engine will produce a levitation force that will give it mobility.

8. How This Energy Will Be Captured

This energy is magnetic in nature. It is known that magnetic energy interacts well with other magnetic fields. Thus magnetic fields are exactly what will be used to capture and contain this magnetic energy.

The material that we know about that will produce the very strong magnetic fields will be superconductors. When Superconductors come in contact with a magnetic field, they then produce and equal magnitude, oppositely directed magnetic field. Using this property, we believe that it will be possible to direct and collect some of the magnetic energy of Superlight. At first, we will use cooled superconducting materials.

As soon as the room temperature materials are developed, the energy generating system will become much less complicated. Our preliminary calculations predict that these energy producing devices will generate a series of high frequency magnetic pulsed fields that will be specifically designed to send radiant magnetic power to whatever device needs it. Such devices for lighting, heating, cooling, cooking, washing or whatever; will each be supplied with the power it needs for its operation, all from the same central power source.

Magnetic energy and strong magnetic fields have been found to be very friendly to the operation of our various body parts. Magnets and magnetic fields are used in many healing devices and applications. Placing magnets directly on the body are currently used by many doctors and practitioners for healing. (Reich, 2014; Payne, 1996; Gerber, 2001) So it is believed that having strong magnetic field generators in our homes, to run all our energy consuming devices, will not be a health problem.

9. Future Applications

Once this power is developed for normal household use, commercial units will be made for all types of industrial applications, and all our industrial energy needs will be supplied without any pollution and at an extremely low cost.

One of the first space applications we will make, will be radios that work on superlight speed and frequencies. Superlight, or Magnetoelectric (ME) energy travels at more than 10 billion times faster than EM radio. To communicate with a person that lives on a planet that orbits the nearest star (8 light years away) becomes practical.

At superlight speed, it is only 3 miliseconds round trip for a communication cycle. As another example of how fast superlight is, a typical galaxy is about 100,000 light years wide from one side to the other, that's pretty far if you want to send a message from one side to the other with EM, but with a Superlight radio it is only 3 minutes.

When we build our star ships we will never ever have to take any fuel will us since the Sea of Energy is all prevailing in space everywhere in our universe. I believe that God created this universe so that His Sentient Beings will be able to travel through the whole universe and spread his life everywhere. That is a great thought to end this article!

Figure 1: A spiral galaxy showing Superlight jetting from both ends of its rotational axis. From NASA, credit: Alfred Kamajian.

Figure 2: Eleven photos of spiral galaxies each jetting out Superlight energy from their rotational axis.

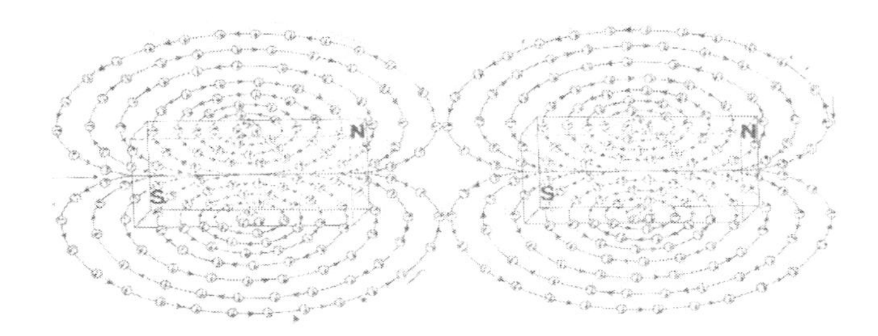

Figure 3: Joseph Newman's concept of the particles creating a "magnetic field."

Figure 4: A Deep space photo taken by the Hubble telescope for 8 hours in ¼ arc sec square area showing about 5000 galaxies. When this number is projected to account for the full spherical sky it calculates to about 400 billion galax-ies.

Comments by Dr. Niels Winsor:

After several years of discussions with Dr. Milewski, it has become apparent that he uses unconventional methods to analyze problems that are not well-handled by conventional theory. In the document above, he constructs a theory of Zero-Point Energy (ZPE) that seems completely outside mainstream scientific theory. However, I have found references in the (mostly) refereed literature that support his model and conclusions. Furthermore, if he is correct in his analysis, the application of this model may lead to significant advances in many areas of science.

10. Background

I received a very conventional scientific education (S.B. MIT, math and physics; A.M. Dartmouth, physics and astronomy; PhD Princeton, astrophysical sciences). However, after discussing the subjects presented above, and many other things with John Milewski over several years, I have come to recognize the shortcomings of conventional science and acknowledge that many of the phenomena John is interpreting do not have satisfactory conventional scientific explanations. For example, "zero-point energy" is often invoked to "explain" the phenomenon of Cold Fusion, which was first documented by Pons and Fleischman more than 20 years ago. Conventional scientists "proved" it was not real. However, those original Cold Fusion experiments have now been replicated more than 100 times, and recent work has greatly extended the production of "anomalous" energy from related devices.

11. Recent developments in Cold Fusion

The field has progressed so far that three companies are now in the process of commercializing four technologies based on these "zero-point energy" devices. Two of these have presented their results at the recent International Conference on Condensed Matter Nuclear Science in Missouri. (Biberian, 2013) The summary of the conference (Nagel, 2013) includes a graph that shows the measured performance of two of these devices on a "Ragone" plot of power per unit mass versus stored energy per unit mass. Using the data from a graph in the conference summary, I have added these two points to a plot that includes nuclear power.

[Please see Figure 5 below.]

Here "eCat" is the Rossi device from Italy, and NANOR is from Jet Energy Inc. in the U.S. Note that the horizontal and vertical axes are both logarithmic. The energy content of these devices is literally orders of magnitude beyond anything that can be obtained from batteries, flywheels, gasoline or hydrogen. Here are Cold Fusion experimental results, put on a plot that includes plutonium, the nuclear power supply that is used for deep space missions. Plutonium is the densest nuclear energy store available to present-day conventional science. This plot shows that whatever these Cold Fusion devices use as a source, in both energy density and power density, it is comparable with what we now consider nuclear energy. This suggests that these devices may draw from the same source, and that our present understanding of "nuclear power" may be missing something important.

John has offered an explanation of that "something missing." I will relate his suggestions to some experiments performed with these Cold Fusion devices, and separate research performed by Tom van Flandern.

12. Gravity and Inertia

First, consider gravity, and a related subject, inertia. I was taught that they are inherent properties of matter, inseparable from it. Gravity is expressed in a "gravitational field." Inertia is a property of matter that causes it to resist changes in its velocity when it is subjected to a force. If you have been reading the scientific literature lately, you may have noticed claims that the Higgs Boson has been discovered, and that it is the particle in matter that gives it "mass." That is, the presence of the Higgs Boson causes inertia and gravity in matter.

Is that the only explanation? No. There is a chain of theories and papers by Sakharov (1968) and Puthoff (1989), to Haisch (1994), who mathematically derived inertia and gravity from zero point fluctuations. Later work by Puthoff (2010) provides an explanation for General Relativity based on this model, without requiring Einstein's curved-space Theory of General Relativity.

Even though this research provides new insight into old things we thought we understood, they still fail to explain other things. The Higgs Boson is a "hot topic" intended to "explain" mass, gravity and inertia. Other "hot topics" in science are "Dark Matter" and "Dark Energy." Why are they needed? The answer is that the theory of gravity, as conventional scientists understand it, cannot explain the dynamics of galaxies without "fudge factors." Specifically, our telescopes and methods have become so good that we can accurately estimate the mass of visible stars in many galaxies. We can also measure the velocities of the orbiting stars outside the galactic center. These results are inconsistent with both classical (Newtonian) gravitation theory and general relativity. Dark Matter "reconciles" these inconsistencies with an arbitrary fudge factor.

Tom Van Flandern (1998) has an elegant explanation for this, and a really fundamental observation, from his student days. This is his observation (Van Flandern, 1998, p. 1):

> *The most amazing thing I was taught as a graduate student of celestial mechanics at Yale in the 1960s was that all gravitational interactions between bodies in all dynamical systems had to be taken as instantaneous.*
>
> *This leads to the unavoidable conclusion that, whatever the mechanism of gravity may be, it cannot obey the "laws" of relativity. Specifically, it must not be limited to the speed of light.*

Once he had this insight, Van Flandern (2008) reviewed the astronomical observations that could put a limit on the speed of gravity, and stated: "Conclusion: the speed of gravity is $\geq 2 \times 10^{10}$ C." He then went further and derived a formula for what he called "the Le Sagian universal law of gravitation." Unike

Newton's expression, for distances compared to galactic dimensions, it has an exponential fall-off of the "gravitational force" and an additional term when a test particle is considered in the field. The detailed formula is in the Appendix.

13. Superlight Time Scales

Milewski has observed that superlight signals would make communication (and possibly transportation) on galactic scales take place at a rate that is easily comprehensible by humans. A figure gives some examples of the times involved.

[Please see Figure 6 below.]

Note that these are lower bounds on the distance gravity can travel in these time units. Van Flandern was only able to set a lower bound on the speed of gravity, based on cosmological measurements.

14. Conclusions

What can we learn from these observations? First, some things (gravitons), which are going much faster than light, can explain gravity on all distance scales. Second, observed zero point (electromagnetic) fluctuations can be shown to explain both gravity and inertia at short distances. Third, wave-particle duality is an established fact for electromagnetic waves. In my view, John's conjecture that superlight is produced by the galaxies and can play the part of these gravitons neatly fits what is known about these effects. If all this can be independently confirmed, I believe it is reasonable to look for experimental confirmation of the rest of his theories on this subject!

15. References

Biberian, Jean-Paul (2013) ICCF Proceedings (in press) http://iccf18.research.missouri.edu/ accessed 11 July 2015.

Coats, Callium (1996) Living Energies Vickor Schauberger' Brilliant Work with Natural Energy Explained ISBN 0-946551-97-9

Callahan, Philip S. (1980) Ancient Mysteries, Modern Visions. The Magnetic Life of Agriculture. ISBN 0-911311

Gerber, Richard, M.D. (2001) Vibrational Medicine ISBN 1-879181-54-4

Haisch, B. (1994) "Inertia as a zero-point-field Lorentz force" Phys. Rev. A 49 (2), 678–694

Matthew, Edwards R. (2002) Pushing Gravity: New Perspective on Le Sage's Theory of Gravitation. ISBN 0-0683689-7-2

Milewski, John V (2009) http://wwwsubtleenergy.com/ormus/tw/superlight.pdf, retrieved 27 January 2014

Moray, John E. (2002) The Sea of Energy in Which the Earth Floats (Salt Lake City: T. Henry Moray Foundation) ISBN 978-1-4691-5980-5

Nagel, Dave (2013) http://infinite-energy.com/images/pdfs/NagelICCF18.pdf accessed 11 July 2015.

Naudin, J.L. (2014) Magnetic Energy to Heal the Planet, http://www.magneticenergy.org.uk/ accessed 11 July 2015

Newman, Joseph (1998) The Energy Machine of Joseph Newman, 4th Ed., ISBN 0-9613835

Payne, Buryl (1996) The Body Magnetic ISBN 0-9628569-9-1

Puthoff, H.E. (1989) "Gravity as a zero-point-fluctuation force" Physical Review A 39 (5), 2333.

Puthoff, H.E. et al. (2005) "Levi-Civita Effect in the Polarizable Vacuum Representation (PV) Representation of General Relativity" Gen. Rel. and Grav. 37, 483-489

Reich, Willham Orgone Energy http://orgone.org/ retrieved 11 July 2015

Sakharov, A. (1968) "Vacuum Quantum Fluctuations in Curved Space and the Theory of Gravitation" Soviet Physics - Doklady 12 (11), 1040.

Tiller, William A. (1975) "Positive and Negative Space Time Frames As Conjugate Systems" Proceedings of ARE Symposium AZ January

Van Flandern, T. (1998) "The Speed of Gravity -- What the Experiments Say" Physics Letters A 250, 1

Van Flandern, Tom (1999) http://metaresearch.org/cosmology/gravity/speed_limit.asp retrieved 11 July 2015.

Van Flandern, Tom (2002) in Matthew (2002), pp. 93-122.

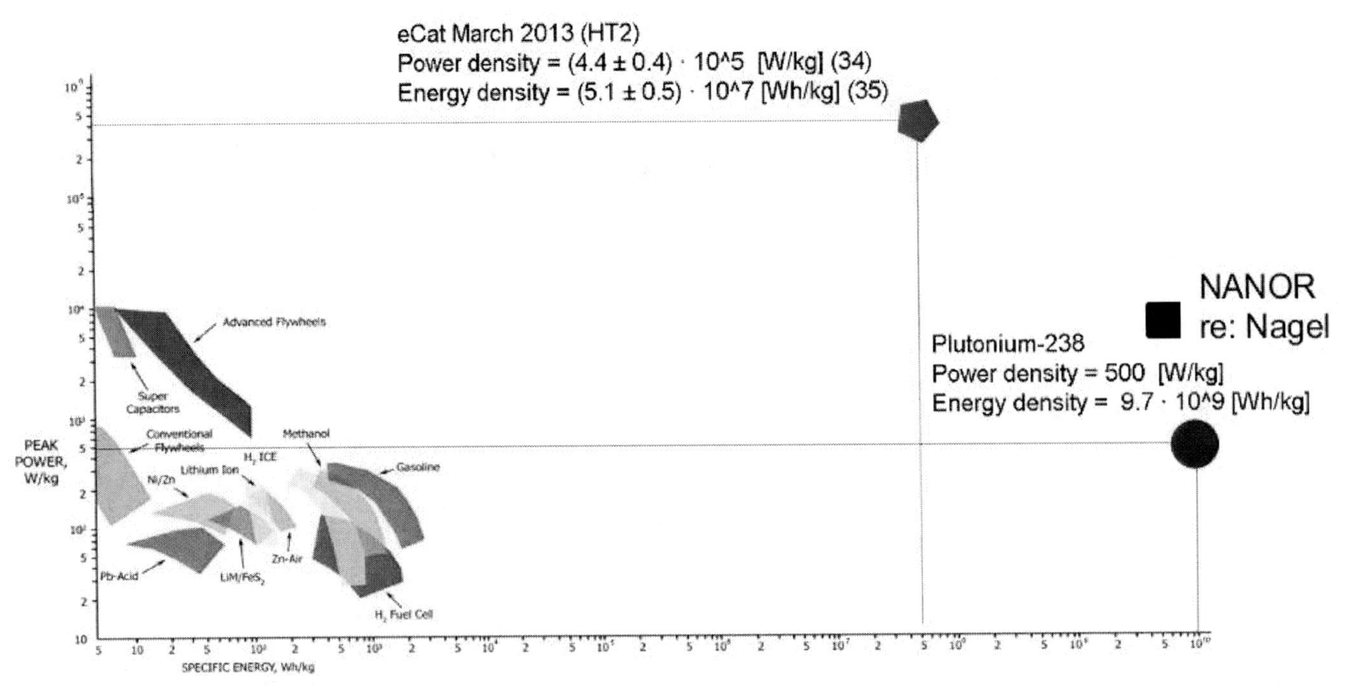

Figure 5: Ragone plot of conventional (bottom left corner), nuclear (black square) energy stores and power supplies plus eCat (red pentagon) and NANOR (blue circle) Cold Fusion de-vices.

Some interesting distances, all in meters

Quantities	Name	Meters	Also
	Light-second	3.00E+08	
L-y	Light-year	2.59E+013	
Pc	Parsecs	3.09E+016	3.26 L-y
MPc	MegaPc	3.09E+022	1e6 Pc
GPc	GigaPc	3.09E+025	1e3 MPc
	Diameter of Milky Way	1.04E+021	3 G-m
	Diameter of Universe	8.95E+026	4.73 G-y
	Gravity-second >	6.00E+018	
G-m	Gravity-minute >	3.60E+020	
	Gravity-day >	5.18E+023	
G-y	Gravity-year >	1.89E+026	
Ratios		Dimensionless	
	Gravity-minute / Light-year >	38025.71	
	Diam. of Univ. / Gravity-year <	4.73	

Figure 6: Comparison of Galactic-scale measurements in light and superlight units. The units are defined in the left column.

Appendix: on Gravity

Tom Van Flandern (died in 2009) produced several important papers on gravity. Two

- **The Speed of Gravity What the Experiments Say** (1998)
- http://metaresearch.org/cosmology/speed_of_gravity.asp
- **The Speed of Gravity - Repeal of the Speed Limit** (2002)
- http://metaresearch.org/cosmology/gravity/speed_limit.asp

are available on the internet. The first of these was published in a refereed journal

T. Van Flandern, **Phys. Lett. A 250** (1998) 1-11. DOI: 10.1016/S0375-9601(98)00650-1
In 2002, Matthew R. Edwards published a book on related theories (Matthew, 2002)

Pushing Gravity: New Perspectives on Le Sage's Theory of Gravitation
http://www.amazon.com/dp/0968368972/ $6.95 as of 11 July 2015.

Several authors of other chapters in that book refer to the Van Flandern work. Van Flandern gives an explicit formula for the "force" of gravity in his chapter:

- The Le Sagian universal law of gravitation, as refined by Tom Van Flandern:

$$\ddot{\vec{r}} = -\frac{GM}{r^3}\vec{r}e^{-(r/r_G)-S_G\int \rho\, dr} - \left(\frac{3\nu\dot{\nu}}{c^2} + d_G\right)\vec{\nu}$$

- where
  - r_G is the rms distance between graviton collisions, estimated to be $1\,kpc \approx 3.09 \times 10^{19}$ m.
  - S_G is the graviton shielding coefficient, estimated to be 2×10^{-19} m^2/kg.
  - $\vec{\nu}$ is the velocity of a test particle (e.g. a star in a galaxy).
  - d_G is the graviton drag coefficient $= 8\pi G\rho_B/3H_0 = \Omega_B H_0$; where G is the universal gravitational constant $\approx 6.672 \times 10^{-11}$ m^3/s^2/kg, ρ_B is the average baryonic matter density of the universe $\approx 3.0 \times 10^{-28}$ kg/m^3, H_0 is the Hubble constant ≈ 60 km/s/Mpc $\approx 1.94 \times 10^{18}/s$, and Ω_B is the ratio of the baryonic density of the universe to the critical density ≈ 0.044. From these values, d_G is estimated to be $\approx 8.5 \times 10^{-20}/s$.
- From experiments, the speed of propagation of gravity is
-
  - $C_{graviton} > 2 \times 10^{10} C_{light}$.

We suggest this gravitational force is the result of very fast particles (Superlight) that weakly interact with matter. According to Wikipedia, this theory was originally proposed by Nicolas Fatio de Duillier in 1690 and later by Georges-Louis Le Sage in 1748. Le Sage proposed this theory 61 years after Isaac Newton published his theory of gravity. The Pushing Gravity book develops the history and evolution of Le Sage's concept, and adds current research by many scientists.

If you find this hard to believe, I suggest reading a book for the general reader by Rupert Sheldrake:

Science Set Free: 10 Paths to New Discovery http://www.amazon.com/dp/0770436706/

Pay special attention to his section How constant are the "fundamental constants"? in Chapter 3.

See also: https://www.youtube.com/watch?v=0QKve_cw82k (accessed 11 July 2015) Superlight Part 1-Antennas, and other internet links to the work of Dr. John Milewski.

Light without Energy

John-Erik Persson
Fastlagsvägen 2, 12648 Hägersten, Sweden
e-mail: john.erik.persson@gmail.com

This article presents alternative interpretations to the empirical evidences behind the theory of relativity and the quantum theory of light. These interpretations indicate the existence of an ether. It is demonstrated that the absurdities existing in modern physics can be eliminated. However, the implied ether must not only exist; the ether must have very remarkable properties. It must have its own state of aggregation and light does not transfer energy.

Keywords: Theory of relativity, quantum physics, atomic clocks, Pioneer anomaly.

1. Background

In quantum physics a wave property, f in hf, is said to prove light to be particles. It is also stated that photon particles colliding with electron particles are said to cause the electrons to move in the direction from which the photon was coming. In the theory of special relativity photon particles are said to move with the same speed c in relation to all not accelerated observers and space and time are regarded as elastic concepts. In the theory of general relativity gravity is explained as the bending of nothing.

These inconsistencies are well known and will not be directly addressed here. Instead the empirical background to this so called modern physics will be discussed. Alternative interpretations will be presented on many points. It is demonstrated that the absurdities can be eliminated by introducing an ether. The ether is a concept that seems to be eliminated due to *lack* of knowledge. This is bad logic. However, as we will see, the ether suggested here must have remarkable properties.

2. Quantum Physics and Ether

The law of energy conservation is the starting point for quantum physics. An electron in a state of constant energy cannot radiate energy. According to experience absorption and emission of light is related to energy change in charged particles. It is therefore concluded that this exchange of energy is with light. Light is therefore assumed to transfer that energy. Contribution of energy from, or to, the ether is not regarded. If we include the contribution from the ether it is possible that the contribution from light is zero. This idea can be united with the fact that we have experience only from absorption and emission but not directly from *transfer* of light. It is therefore possible that light does not contain energy. Light can be considered to contribute information only to a process where energy is exchanged between electron and ether. Information without energy can for instance be represented by polarization of ether particles. Energy is needed only when polarization is changed but not for maintaining constant polarization. Energy must not necessarily be transferred by light. Light is a field without energy and transfer of energy starts at first when an electron is placed in this field. Light propagation over billions of light-years and destructive superposition can be united with the idea of no energy in light. If these ether particles have mass, gravity and energy can be explained by the same ether particles.

Assuming light without energy means that bound electrons can radiate continuously without losing energy. We do not need quanta of light. Electrons, as well as kernels, can contribute to blackbody radiation. Atomic kernels can dominate at lower frequencies, and lighter electrons can dominate in the high frequency range. The fast decrease in radiation intensity, at higher frequencies, can be caused by light frequencies equal to integer multiples of electron's orbiting frequency.

The common explanation to the photoelectric effect is based on a light particle colliding with a loosely bound electron particle. Kinetic energy is supposed to be exchanged. Light particles moving towards a crystal are assumed to force an electron to move *away* from the crystal. This assumption is not realistic in relation to the laws of mechanics. Instead, we can find a more logic model based on the wave interpretation of light. The inertia from electron mass is compensated by a Coulomb force from electron charge. These forces are transverse to motion. Light with a frequency equal to (or to an integer multiple of) the electron's orbiting frequency produce a force transverse to motion and can make interference with the two balancing forces. Due to this interference, the disturbing force can be integrated over many orbiting periods. This means that *waves* of light can change the *potential* energy of the electron without large changes in kinetic energy. When potential energy is changed to about zero the electron can escape its kernel with about the same kinetic energy as it had before the interference. This means that the electron must have a certain amount of kinetic energy before the interference to allow emission (or escaping). The electron must be *tightly* bound. The frequency dependency in kinetic energy in emitted electrons is thereby explained by this demand on high initial kinetic energy. Quantization is therefore not needed.

A remarkable fact in the common interpretation of the photoelectric effect is that a *wave* property, f in $\Delta E=hf$, is said to prove light to be *particles*. In this article h is instead considered as a scale factor only. The assumption of interference demands light frequency to be an integer multiple of the electron's orbiting frequency. This fact can explain the very fast decrease in radiation intensity at higher frequencies in the blackbody radiation. When the contribution from kernels is very low in blackbody radiation it is observed that the electrons produce a sharp line type spectrum. This fact supports the idea that bound electrons in atoms radiate blackbody radiation.

X-rays appear to be generated when a fast electron is captured by an atom. X-rays are then generated for a short period of time. A process in opposite direction is also possible. This means that X-ray waves interact with a *bound* electron in the same way as light in the photoelectric effect. This means an interference phenomenon. When the electron's potential energy is near zero the electron can escape its kernel with about unchanged kinetic energy. However, the electron can be captured by another atom. This capturing can generate a new X-ray wave packet with somewhat different properties in relation to the first wave packet. Since escaping and capturing is done in different atoms such differences are possible. They can be different in direction, frequency and time duration. This process is called Compton effect but is in reality two processes. The involved electron escapes one atom but is captured by another one. The Compton effect is therefore best described by the wave model for light and interaction with a *bound* electron.

By assuming light without energy we can abolish many quantum paradoxes. However, there is a cost for this simplification and it is not enough to introduce an ether. We must also ascribe remarkable properties to this ether. The ether is not solid, liquid or gas, but must have its own state of (no) aggregation. Ether particles do not collide with each other, and the ether must be super fluid. Accepting electrons to radiate without losing energy can perhaps help us to understand why planets can be orbiting without losing energy. The planets can generate a wave function that hides ether wind around our planet from observers on our planet, just like how electrons generate blackbody radiation. Near our planet we have instead a vertical ether wind explaining gravity. Maintaining the wave function demands no energy but we need energy to change the wave function. This fact can explain inertia.

If we accept light without energy and super fluid ether we can abolish the wave or particle confusion as well. Light is waves and ether is particles. Destructive superposition in light can be explained. All we have to do is to accept light waves to interact with *bound* electrons in the photoelectric effect and in the Compton effect.

3. The Three Directions of Light

Light is transverse oscillations in two dimensions inside a wave front. This is demonstrated by the concept polarization in light. The state of motion of the ether defines the reference for the constant wave velocity c. The propagation of transverse light waves is different from (and more complex than) the propagation of longitudinal sound waves. It is therefore not correct to use sound waves as a model for light waves. We must remember that c and v are very different concepts. v is a real motion of matter but c is a motion of behavior only. c and v also differ in many orders of magnitude in strength. If we use light that is focused into a very narrow beam and detect the direction of that beam we find that the vector sum **c**+**v** describes this *real* motion of light very well. In this case we use not coherent detection.

However, the direction of the center of a focused beam is equal to the direction of the normal to the wave fronts *only* in the frame of the ether. The beam direction is changed by a changing transverse ether wind but the wave front orientations inside that beam are *not* changed. This follows from the fact that the ether wind in transverse direction has the same effect in *all* points on the wave front. The wave velocity c has also the same definition in all points on the wave front.

In an interferometer we have mirrors in the interferometer and in laser cavities that define the wave fronts to be parallel to these mirrors. Wave fronts are defined by mirror orientation *independent of mirror motion inside mirror plane.* Moving the mirrors inside their own plane does not change direction of light but means only that light after a round trip (in Michelson's interferometer) hits a different point (in relation to starting point) on the mirrors. This fact is normally not observable in an interferometer. Interferometers are sensitive in one dimension only and transverse mirror motion (or ether wind) cannot be detected. Therefore, in an interferometer, light's wave vector **c** is always orthogonal to mirrors independent of transverse ether wind. This means that in optical experiments where mirror feedback defines wave fronts we always get the same value c transverse to mirrors. Interferometers are *blind* to transverse ether wind. The irrelevance of transverse ether wind in interferometers means that relevant description of light is $\mathbf{c}(1+v_c/c)$ in this case (v_c is component in **v** parallel to **c**).

The irrelevance of transverse ether wind means that Stokes was wrong when he reduced Michelson's prediction by 50 % due to transverse ether wind. Stokes used Pythagoras theorem in a wrong way and his effect does not exist. Einstein was also wrong when he assumed the effect real but hidden by something he called dilation of time. Instead time dilation does not exist either. It is demonstrated in Fig 1 that unchanged boundary conditions mean unchanged light behavior.

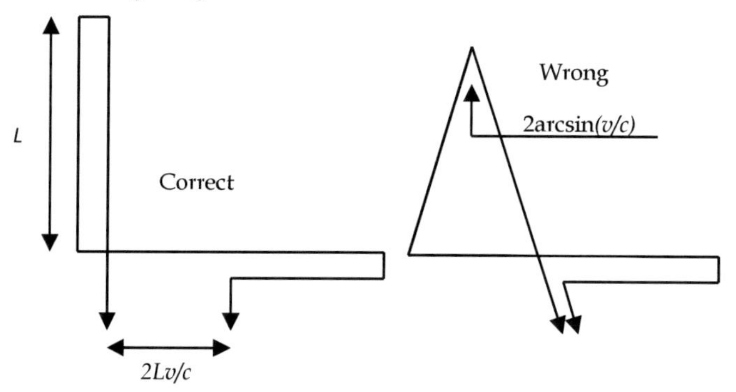

Fig 1. Michelson and Morley's Experiment. Wave fronts defined by mirrors in laser cavities and in interferometers do not change when the equipment is moving inside mirror's plane. The orientations of mirrors still define light to move with the speed c transverse to mirrors. Instead speed in relation to equipment is changed to $(c^2+v^2)^{1/2}$.

The fact that the waving is controlled by boundary conditions implied by mirrors means that we have different meanings in the concept direction of light. When a focused beam is detected not coherently we can see the *real motion* of light that is defined by **c** and **v** together. However, in interferometers we only have possibility to observe $\mathbf{c}(1+v_c/c)$ and direction is defined by **c** only to be in a direction transverse to mirrors. We can only see *wave motion* and only *one* component in the ether wind is relevant. In telescopes we have a third meaning in the concept direction of light. Telescopes detect also the wave orientation independent of **v** but changes in the detector motion **u** transverse to light produce an aberrated and false wave front orientation. The effect is observed in stellar aberration. The transverse component u_T in **u** changes apparent direction to a fix star an amount equal to arctg(u_T/c). We have therefore a third meaning in the concept light direction as *apparent wave motion*. This direction depends on **c** and **u**.

This explanation was provided by Bradley for light as particles. The effect of observer motion must be the same for light as waves. Both phenomena move with the speed c along a straight line. This is valid for light as long as we consider light direction as the normal to the wave fronts and transverse ether wind is irrelevant. Therefore, stellar aberration reflects our own state of motion. Stellar aberration cannot tell us anything about the ether wind due to irrelevance of transverse ether wind. This is described in Fig 2. We can therefore conclude that the concept direction of light can have three different meanings. Real motion dependent on **c** and **v**, wave motion dependent on **c** and apparent wave motion dependent on **c** and **u**.

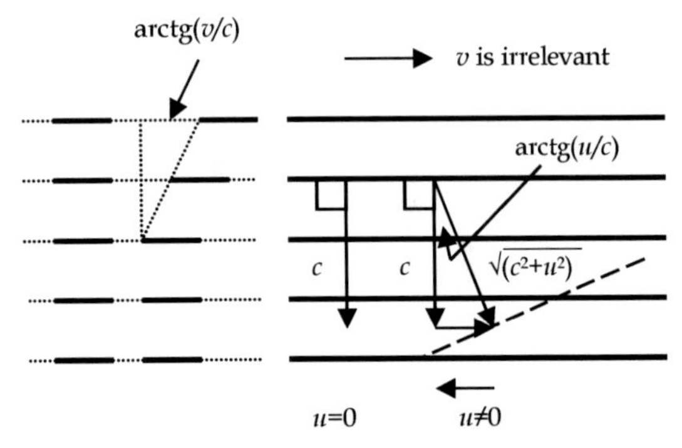

Fig 2. Stellar aberration. The left diagram demonstrates how vector **v** is relevant in focused light. The right diagram demonstrates how transverse ether wind becomes irrelevant if light is not focused. We can also see how observer motion **u** produces an aberrated (and false) wave front orientation.

4. Gravity and Ether Wind

The mass of our planet generates a spherically symmetric property in the ether. This property is called gravity and the effect of gravity can be explained by an attenuation of a flow of ether particles passing through our planet. The ether, in itself, is not entrained but the gravity property is entrained. Gravity should not be described as a bending of nothing, but rather as a property of the ether.

The high stability in the orbits of planets around our sun demonstrates no aberration in gravity. The force of gravity is directed exactly towards our sun. This fact has been compared to the aberration in light from our sun and generated the idea that gravity propagates with an enormously high speed. This idea can be wrong since gravity can be considered as stationary (in relation to its source) in space. Lack of motion can explain lack of aberration. Observer motion is relevant in relation to moving light but not in relation to stationary gravity. The speed of ether particles can be relevant in relation to *changes* in gravity but not in relation to constant gravity.

Gravity theories based on particles moving towards, or from, a massive body have been suggested. However, the most interesting model is based on particles moving in *all* directions. Such a model was suggested by Fatio, and was further developed by Le Sage. This model contains a cause of gravity, since an attenuation of ether particle flow is caused by matter. The attenuation produces an asymmetry in the flow of particles. An ether wind is produced and this ether wind produces gravity. The number of particles leaving a body is reduced in relation to the number of approaching particles. The magnitude of the generated ether wind is many orders of magnitude smaller than the speed of individual ether particles. This is valid for bodies of the size of bodies in our planetary system.

The value of the vertical ether wind, causing gravity, is not known. However, a preliminary hypothesis was made in [1]. It was assumed in that article that the magnitude of the vertical ether wind near our planet was equal to the speed of a satellite in a circular orbit around our planet at the same altitude as the ether wind. This hypothesis was also used, in the same article, to estimate different phenomena.

Ether theories are usually separated into three categories, namely autonomous, entrained and nonexistent. The theory presented here belongs to none of these. It is instead assumed here that it is only properties that are entrained. These properties are an ether wind defining the reference for the constant wave velocity of light c, and the field of gravity caused by this ether wind. However, the presence of mass not only generates an ether wind directed towards the center of this mass, but also hides the surrounding ether wind from observers near this mass. Therefore, we cannot detect the speed due to our planet's motion in relation to our sun. Celestial bodies can be assumed to generate a wave function in the ether. Energy is not needed to maintain this wave function, but energy is needed to change this wave function, which can explain inertia.

It was earlier stated that charged particles interchange energy with the ether. Not with light. Light contributes with information only. The information can be represented by polarization. However, these ideas can only be possible if we can assume remarkable and super fluid properties of the ether. On the other side there are many advantages with such a theory since lots of absurdities and paradoxes can be abolished. Light without energy has much to offer.

Newton and Einstein describe the *effect* of gravity by existence of mass. The Fatio and Le Sage model explains also the cause of gravity by attenuation (or shielding) of existing ether flow by matter. Ether particles are moving in all directions and the shielding effect produces an asymmetry in the flow that causes the ether wind and thereby gravity. The produced force has a very small and not linear effect in Le Sage's gravity that is not present in Newton's gravity. This difference can be exemplified in a homogenous sphere. Newton predicts a linear function of radius but according to Le Sage we should expect an exponential function approaching an upper limit. This limit is very high and the difference between the two theories is therefore very small, for bodies of the size of our planet.

During a solar eclipse a small decrease related to the mentioned nonlinearity in the combined gravity from Sun and Moon should produce a very small increase in vertical gravity as indicated in a very sensitive gravimeter. This effect has been very roughly estimated to be somewhere between 10^{-9} and 10^{-8} in relation to gravity on Earth. See [1]. An effect of about this magnitude was detected in China in 1997 during a solar eclipse. See [2]. The registration was about zero in the middle of the eclipse with two bumps with 'wrong' sign just before and after the eclipse. This phenomenon is explained by the notion that this problem is a four body problem. The gravimeter detects not only the effect in the gravimeter but detects instead the difference between effect on gravimeter and effect on nearby parts on our planet. These two effects are about equal in the middle of the eclipse but the effect on our planet dominates just before and after the eclipse. This explains the two bumps. Observations of the same kind have been made in high TV-towers. Horizontal motions during eclipses have also been observed to demonstrate effects before and after the eclipse of opposite sign in relation to the effect in the middle of the eclipse. The idea that we have a four body problem is supported.

The shielding effect has been studied by Majorama who used the term 'apparent mass' to explain the shielding effect. Majorama's ideas seems interesting but his empirical results appears to be much larger than estimations by this author.

5. First Order Effects of the Ether Wind

The first order effect of motion in relation to the ether was detected by Sagnac [3]. He demonstrated a translational effect in light propagating along four straight lines. Translational effect has also been observed in the global positioning system (GPS). The effect is observed when clocks in two time stations on Earth are compared. Compensation for the rotation of our planet must therefore be done in GPS. When two clocks are separated a length L and an ether wind v is blowing in the direction of L we must compensate for a change in propagation time L/c by a factor $1\pm\beta$ ($\beta=v/c$). The size of this compensation is still the same even if the comparison is done over a not straight line. The effect detected in four straight lines by Sagnac can today be detected in one straight line. Dr C C Su has described how this can be done. See [4] and [5]. Two HeNe lasers with good frequency stability are connected to an interferometer. They are separated a couple of meters and mounted on a platform with very high mechanical stability. By changing direction of measurement vertical or horizontal ether wind can be detected.

Stellar aberration has been regarded as relevant for the ether's state of motion. We have earlier seen that transverse ether wind is irrelevant in telescopes and that instead telescopes transverse motion is relevant. Stellar aberration reflects therefore only observer motion in agreement to Bradley's interpretation [6]. Stellar aberration is useless in relation to the ether wind.

Another manifestation of a first order effect of the ether wind is observed in the bending of light near our sun. The wave motion of light depends only on the longitudinal component in the ether wind. This component is first positive and later negative for light tangential to the Sun. This is an effect of the falling ether near the Sun. Since the effect is largest nearest to the Sun. we get a bending first away from the Sun and later back to the same direction. The bending is not to the same position. The difference can be calculated by integrating the gradient in longitudinal ether wind along the path of light. This calculation has not been done but a very rough estimation has been done in [1]. This estimation gave an apparent change in direction of 10^{-5} radians. This is in agreement to observation. A more accurate calculation should be done.

6. Second Order Effects of the Ether Wind

The separations between atoms in a crystal are controlled by fields in the ether generated by the atoms. Changes in these fields are propagated with the speed c in relation to the ether. When the crystal is moving with the speed v these changes propagate with the speed $c(1\pm\beta)$ in relation to the atoms ($\beta=v/c$). The ether is the only possible medium for this interaction. In this way positional information is transferred between atoms. Two nearby atoms are in a two-way communication based on the ether. The ether wind has opposite effects in two opposite directions but a very small difference means that a second order effect is produced. The separation is reduced by a factor $1-\beta^2$ in the same way as the reduction of two-way speed of light. Michelson also used two-way communication based on the ether in his measurements with light together with Morley [7]. The only difference is that Michelson used *sequential* communication and atomic separation is based on *simultaneous* communication. It is therefore reasonable to assume that the reduction in two-way light speed is compensated by an equal effect in atomic separation. This means that Michelson's method is *useless* in relation to the ether wind. The autonomous ether is therefore not refuted by Michelson's experiments.

Bound electrons are orbiting a kernel in atomic clocks. The electron's inertial force is balanced by a Coulomb force. Changes in the Coulomb force field can be propagated with the speed c from kernel to electron. An ether wind β ($=v/c$) means that the Coulomb force appears to emanate from a point β times radius behind the kernel. However, the inertial force is also changed in the same way since the form of the field is still a circle. Therefore we get no accelerating effect of β. Or?

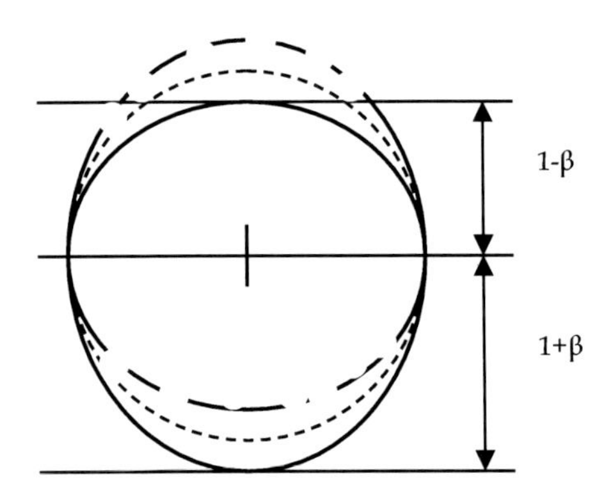

Fig 3 The effect of the ether wind β on the Coulomb force field. Only one dimension in the field is changed by the ether wind. In front of the kernel the field is a compressed ellipse in the electron's orbiting plane and behind we get an extended ellipse.

The simple explanation above does not regard the ether properly. If we instead assume the existing ether correctly we can conclude that the Coulomb force is changed by ether wind in one dimension only. Light is a wave motion. Therefore the *form* of the Coulomb field is changed. The field is not a eccentric circle but two half ellipses in the plane of orbit. In front of the kernel the field is compressed by a factor of $1-\beta$ and behind it is extended by $1+\beta$. No changes are caused in the points beside the kernel. See Fig 3. When the electron adapts to this field the electron is accelerated and decelerated by β in the direction of the ether wind. The effect of this acceleration and deceleration is that electron speed is changed between $w(1-\beta)$ in front of the kernel and $w(1+\beta)$ behind the kernel and w beside the kernel. Therefore the electron is moving with the speed adjusted by $1\pm\beta$ transverse to the ether wind. Speed proportional to $1\pm\beta$ means time proportional to $(1\pm\beta)^{-1}$ for the same distance, that is a half period. This means that the time for a full period is changed in proportion to $(1-\beta^2)^{-1}$. The frequency of an atomic clock is therefore proportional to $1-\beta^2$ with β as the component of the ether wind that is falling inside the plane of electron's orbit.

Blackbody radiation, photoelectric effect and Compton effect have all been explained by the wave model for light. It is therefore logical to explain the structure of the Coulomb field by the wave model for light. When **v** is orthogonal to **c** we find that v is irrelevant for the *wave motion*. Light waves move from kernel to electron with speed c independent of v. The Coulomb field is therefore changed $1\pm\beta$ in one dimension only. We get two half ellipses in the plane of the electron's orbit. We can conclude that we have a slowing of electrons moving forth and back in relation to the ether wind that is of second order in β. This electron behavior is therefore comparable to light behavior in Michelson and Morley's interferometer experiment. This means that we can explain a phenomenon observed in atomic clocks in the global positioning system by means of one mechanical process inside the clocks. In the theory of relativity this behavior is explained by two metaphysical properties of the time concept. These properties (called dilation of time) are not needed in the theory considered here.

By using the assumption about a vertical ether wind equal to the speed of a satellite in a circular orbit on the same altitude as the ether wind we can calculate the change in clock speed when a GPS satellite is put into orbit. Frequency proportional to $1-\beta^2$ is assumed. The satellite is assumed to be stabilized in the direction towards Earth. Clocks are assumed to be orthogonal to this direction. Due to lack of stabilization in the direction of motion the horizontal component is reduced by a factor 0.5. This factor represents the average value of a squared sine function. On ground the assumed vertical ether wind is 7.91 km/s (60 μs/day) and horizontal (due to planetary rotation) in the order of 0.3 km/s (≈0.1 μs/day). In a satellite we have a vertical and horizontal ether wind of 3.87 km/s (14.3 μs/day). With the clock orientation orthogonal to the direction to the Earth we get 60-14.3-0.5(14.3-0.1)=<u>38.6 μs/day</u>. Increased speed. If the clock instead is oriented along the direction towards Earth we get instead 0-(14.3-0.1)=-<u>14.2 μs/day</u>. Decreased speed. The contribution from vertical ether wind is zero, but the contribution from horizontal effect is higher, since satellite rotation is irrelevant, and motion is always inside the electron's orbit.

Two spacecrafts, Pioneer 10 and 11, have demonstrated a very small reduction of speed in their motion in a radial direction out from our solar system. This represents a very small anomaly in relation to modern theories in physics. This phenomenon has been discussed for many years but has still not got a clear explanation. The effect has been detected by means of high precision, two-way Doppler measurements. A new interpretation is presented in this article. This new interpretation is based on the assumption that apparent deviation is only an illusion. Instead of a decrease Δu in Pioneer speed the real effect is an increase Δc_2 in the two-way speed of light. This represents a change in local ether conditions. This is a confirmation of the existence of an ether, but is not in conflict with modern theories of gravity.

In [1] the two-way speed of light was described as $c_2=c\cdot(1-\beta^2)$. ($\beta=v/c$) with v as the ether wind. Horizontal ether wind v is a very small translational effect of the rotation of our planet. In vertical direction the effect is greater and causes the force of gravity. The relation above is in agreement to Michelson's own prediction without the reduction that later was done by Stokes.

The orbiting speed of our planet is $v=10^{-4}\cdot c$. According to earlier given hypothesis we therefore get a two-way speed of light as $c_2=c\cdot(1-10^{-8})$ in radial direction in relation to our sun. This is valid at a distance of 1 AU (astronomical unit) from our sun. At the distance r_{AU} we get $c_2=c\cdot(1-10^{-8}/r_{AU})$. Therefore we can find the change in f as $\Delta f/f=2\Delta u/c=2\Delta c_2/c=2\cdot10^{-8}/r_{AU}$. With a carrier frequency f of $2\cdot10^9$ Hz we get $\Delta f=40/r_{AU}$ Hz. Assuming r_{AU} to change from 20 to 80 AU we finally find that the change of frequency is 1.5 Hz. This result can explain the Pioneer anomaly.

Stokes' reduction of Michelson's prediction by a factor of 0.5 is not correct. Instead Michelson's original prediction is correct but hidden by a contraction of physical objects. Second order effects of the ether wind are instead demonstrated in atomic clocks and in the Pioneer anomaly. See [5].

7. Summary

Quantum physics is based on the law of energy conservation stating that bound electrons in a state of constant energy cannot radiate energy. However, the contribution from the ether to the energy balance is not regarded. Electrons exchanging energy with the ether means that we no longer need interchange with light. Light can be a field without energy in itself and energy interchange starts at first when an electron is placed in the field. We have no evidences stating that energy is transported with light. Light without energy means that bound electrons can radiate blackbody radiation. Light waves can continuously change *potential* energy by interference in electrons and photoemission can be explained. Interference can also explain how X-rays can change potential energy in a similar way and allow an electron to escape its kernel. When the escaped electron is captured by another kernel a new X-ray wave packet is generated and Compton effect is explained by *two* processes. Therefore light can be explained by the wave model only. The notion that the ether can transfer light without energy can perhaps help us to understand how planets can move without losing energy.

The real motion of light is described by a vector sum of wave velocity and ether wind. This motion is relevant when light is observed in a focused beam. The wave motion of light is defined by optical feedback in cavities and interferometers and depends on the ether wind in one dimension only. We cannot observe transverse ether wind in interferometers and only the longitudinal component of the ether wind is therefore relevant. The aberrated wave motion is relevant in telescopes where detection is coherent (as in interferometers) but where observer motion creates an illusion of a changed wave front orientation. Therefore, stellar aberration reflects only observer motion and not ether motion. Stellar aberration cannot rule out the entrained ether.

A theory of gravity described by Le Sage predicts a very small decrease in gravity from the combination of our sun and our moon. An effect in agreement to this idea was observed in China in 1997. See [2]. An effect of the same kind has also been observed in horizontal motions in very high radio-towers.

Sagnac detected a translational effect of the ether wind in four straight lines. In the global positioning system the effect of motion in relation to the ether is observed when time in time stations on our planet are compared. The time for light to travel between two points is changed by a factor $1\pm\beta$. This effect is the same even if the comparison is done over a not straight line. The effect Sagnac detected in four straight lines can to day be detected in one straight line. Dr C C Su has described how this can be done. See [5]. A first order effect of the ether wind is also detected in the bending of light near our sun. This effect is not produced by transverse ether wind. It is produced by the gradient in longitudinal ether wind. See [1].

Michelson failed to detect a second order effect of the ether wind in two-way speed of light between mirrors. The searched effect is compensated by the same effect in the separation of the mirrors. An effect comparable to the one Michelson searched has been observed in atomic clocks. The second order effect is also observed in the Pioneer anomaly. The theory described here predicts a change in frequency of 1.5 Hz when space station is moving between 20 and 80 astronomical units. This is in agreement to observation. See [8].

Light without energy means light without quanta of energy. This means also physics without the wave or particle confusion. Ether is particles and light is waves. If we are brave enough to assume light without energy we can also accept wave functions generated by matter and adapting surrounding ether in such a way as to allow constant motion of planets without loss of energy. However, *changing* this wave function demands energy and explains thereby inertia. The presence of matter produces an ether wind by an attenuation of ether particles. This ether property causes the force of gravity. This means that there is an upper limit on the force of gravity defined by an existing flow of ether particles. We do not need black holes. Cosmological red shift can be explained by falling ether instead of by expanding universe.

8. Discussions

Maxwell shifted light model from particles to waves based on the assumption about an existing ether. This giant step forwards was followed by a step backwards when the same ether was denied. Many years of failures to unite the ether concept with observations was considered as evidences for the ether to be nonexistent. But failures have many reasons and prove therefore nothing. Nonexistence cannot be proved, and is therefore only a conjecture. An important reason to problems in physics is an ambiguity in direction and apparent direction of light. Another difficulty is the fact that **c** is many orders of magnitude larger than **v** and **u**. Scientists have been very stubborn in the denial of the ether concept. In order to exclude the ether lots of absurdities have been included instead. We have got multiple time concepts (twin paradox), multiple universes (universe means everything), black matter (without black ether) and lots of more of this kind.

Instead of basing physics on failures in Michelson and Morley experiments and in stellar aberration we should base physics on successes like atomic clocks, GPS system and Pioneer space station. Atomic clocks and Pioneer space station indicate that Michelson's hypothesis was correct but his test was interpreted in error. First order effects in the GPS clock synchronization and in the bending of light near our sun give also support for the ether's existence. This indicates also that Sagnac's effect in four straight lines also can be demonstrated in one straight line as suggested by Dr C C Su. [3], [4] Therefore, light needs a waving ether and gravity needs a falling ether. Mass and energy are needed in the ether, but not in light. All light behavior can be explained by the wave model. Therefore a Nobel Prize for the particle interpretation of photoemission was unhappy.

It is very easy to test the theories presented here. By changing the orientation of an atomic clock on ground from horizontal to vertical we should increase clock speed by 7×10^{-10} or 60 µs/day. We can also use Dr Su's method to measure the vertical and horizontal ether wind.

9. Conclusions

The concept direction of light is ambiguous and depends on how light is detected. Ignorance of this fact has created the *false* idea that stellar aberration excludes the entrained ether. This error produced the wave or particle confusion with many consequences.

Without an ether electrons were assumed to exchange energy with light instead. This error created the idea that light must be *quantized* into particles.

Without an ether as the reference for the speed of light the observer was used as a reference for light speed. This assumption created the theory of *special* relativity with elastic space and time.

Without an ether as the cause of gravity geometry was used to describe gravity as a bending of nothing. This assumption created the theory of *general* relativity.

With an ether many paradoxes can be eliminated but the ether must have its own state of *aggregation*. Light must be without energy.

With an ether we can explain stellar aberration, Michelson and Morley's experiments, Sagnac effect, bending of light near our sun, slowing of atomic clocks, cause of, and anomalies in, gravity and the Pioneer anomaly.

References

[1] John-Erik Persson,"The Falling Ether", *Proceedings of the Natural Philosophy Alliance 2013,* available at www.worldnpa.org/site/ Search on my name under 'Members'

[2] Qian-Shen Wang, "Precise Measurements of Gravity Variations during a Total Solar Eclipse", Physical review D 62 041101-1.

[3] Georges Sagnac, L'ether lumineux demontre par l'effet du vent relatif d'ether dans un interferometre en rotation uniforme," Comptes Rendus, **v157**, pp. 708-710 (Oct 1913) http://gallica.bnf.fr/ark:/12148/bpt6k31103/f708.table

[4] C. C. Su, J. C. Eur. Phys 21, 701-715 (2001), Digital Object Identifier (DOI) 10.1007/s 100520 100759

[5] John-Erik Persson,"Detecting an Ether-Wind of 10^-6 Based on Sagnac Effect", *General Science Journal* available at http://gsjournal.net/ See my name under 'List of authors'

[6] J. Bradley, Phil. Trans. Roy. Soc. **35**, 637 (1729)

[7] Albert A. Michelson, Edward W. Morley, "On the Relative Motion of the Earth and the Luminiferous Ether". *American Journal of Science* **203**:333-345 (Nov 1887).

[8] John-Erik Persson,"Pioneer Anomaly and the Ether Wind", available at www.worldnpa.org/site/ Search on my name under 'Members'

Structure and Gravity

Paul Schroeder
8244 Anna Ave. Wind Lake, Wi. 53185
e-mail: pshrodr8@aol.co

The details of working with space are well analyzed and understood by astrophysics study but big parts are missing within the overall cosmological views.

1. Introduction

Particle physics is the breaking down of matter into components such as atomic particles. The purpose is to gain understanding of all matter and its place in the universe. Radiation is categorized separately and shows itself essentially via motion. The breaking down of radiation is therefore into lines of flow. Overall field and vortex analyses are insufficient. A proper analysis reveals the nature of gravity, of orbiting and other motions, and reveals the structure of the universe in its totality.

2. Radiation

Radiation has provided the advancements of science in the last 50 - 100 years. It has not been properly assigned its place within our universe structure. Radiation arrives everywhere from all directions. Likewise it comes from all 3 dimensional directions throughout the universe. A point in remote space exists because radiation continuously passes through from all directions. There is no void in the universe. Should one seek void space within a point, that void would vanish to the ongoing radiation flow coming from continuously shifting angles. The structure of the universe is defined by lines of radiation. There is no need for an aether to carry light and EM radiation to provide the structure.

Radiation is called rays as it departs a source and beams as it travels and arrives. What we called a source, such as a lamp or a gravitational body, is but a modifier changing the characteristics of existing radiation. All radiation travels at velocity c depending on its medium as defined within physics. The prominent characteristic of radiation is waves - length and frequency -. Light is the predominant radiation and serves as a central focus of a scale of all possible radiation. The scale, detailed by wave length, describes the various radiation effects which depend on the wave frequency.

As we are dealing with structure, note that EM waves are considered transverse and as beams travel through three dimensional space being viewed as sin waves. That view is two dimensional. Being similar from all viewing directions the waves are therefore actually coils. We may use the terms coil and wave interchangeably.

Radiation has the ability to push matter. This violated early physics laws and led to the duality of wave/particle nature of light. Light is known to push in vision and X-rays are known to push from the Compton Effect studies. Pushing is a transfer of motion and does not depend on light having photon particles.

Since radiation can push it can function as a source for gravity. Thus gravity originates as a push. The attraction nature of gravity is its appearance to us. The attraction occurs when the push in one direction is less than in another. Since a spatial body is involved, the 'net' push becomes apparent as attraction. Magnetism is also a 'net' push of gravity in which, by their spin, flowing electrons can redirect gravity beams coming from one direction.

3. Reduction

Long wave radiation has more potential to penetrate matter than does higher frequency short waves. Long straight beams are like arrows and lack the wave that impacts and stops the flow. The penetrating long wave beams both push masses and are modified by the masses. The modifying action within masses has atomic particles causing two parallel beams to wrap together into one wave without quite doubling the frequency. Wave merging continues throughout and results in fewer beams of higher individual frequency, but there remains less total frequency, and thus less energy. The longer the path of merging, the fewer beams remain to exit. Thus the exiting beams cannot offset the incoming downward beams and a 'net' downward pressure occurs. The pressure/force average is determined relative to the mass center as a reduction process rather than a creation of attraction. The diameter determines the path length. By time of exit the remaining penetrating beams have higher frequencies and exit as radio beams for example from earth and light beams from the sun. The new frequencies depend on the celestial body size.

4. Orbiting

While the reduction process outlined here is unique, the primary revelations of this pushing gravity lie in orbiting, matter creation, and the potential to modify gravity. I will touch on orbiting here. Part of Newton's model is that motion in space continues without change unless impacted by other matter. The concept of prior unchanging motion in space solved the friction issue with space defined as void. We know today that space is not empty. A driving force is needed to explain continuation of orbital motions. For orbital motion that driving force must relate to the center body in some way. Orbiting is revolution relative to the center which is otherwise the relative rotation of the center. All depends on your point of observation.

That center, which rotates relative to an orbital body, is ejecting the diminished gravity beams while pushing them in its rotation circle. Solar beams arrive at an orbiting planet from the right causing pressure toward the left. We accept that planets are pushed toward the sun by a 'net; of the inward gravity. The orbital is also pushed counterclockwise around the sun. The two 'net' force directions, inward and leftward, achieve balance. For comparison, aberration of sun light for observation upon a planet is unlike external starlight aberration. Both light and gravity beams from the sun curve in toward earth avoiding any appearance of earth overtaking the beams. Rotation is the overlooked source motion maintaining the universe.

Pushing gravity beams recycle. Light from distant stars gradually fades out as the wave lengths get longer. The lengthening is caused by the retention pull of the source and the forward pull of the destination body. Light stretches into infrared, microwaves, sound waves and finally gravity waves. The rebuilding is needed as the waves that penetrate matter locally lose some of their energy. That energy loss is the gravity we recognize here on earth and for other planets. The downward push is the 'net' of two opposite pressures.

5. Conclusion

Physics could never work with the gravity source as it was considered a linear pull to an unattainable center. Switching to an external incoming source opens a whole new chapter for modifying the gravitational effects of gravity beams. We can work with it. We can try to block it. The common 'attraction' force of magnetism can be seen as a redirecting of gravity beams by the spinning of matter, ie electrons. There is no reason this can't be done on larger scales. We spin propellers such as helicopter blades. They raise up a big carriage/cockpit attached below. The theory is that the spinning air pushes downward, but isn't it easier to view the motion as being caused by the blades pushing aside the incoming gravity beams. Then the push from below is greater than the push from above. By extension, how did the ancients move the huge blocks forming pyramids and stones such as those at Stonehenge and Easter Island? Wouldn't the blocking of gravity be a logical method?

On the Cause or Causes of Inertia

Duncan W. Shaw
1517 Angus Drive, Vancouver, B.C., Canada, V6J 4H2.
e-mail: duncanshaw@shaw.ca

This article sets out two possible causes of inertia. The first is that a sub-atomic substance - aether - restricts the movement of atoms and atomic matter. The second is that vibrations within atoms restrict the movement of atoms and atomic matter.

1. Introduction

What is inertia? In their treatise, Principles of Mechanics and Dynamics, William Thomson (Lord Kelvin) and Peter Guthrie Tait define inertia: [1]

"Matter has an innate power of resisting external influences, so that everybody, as far as it can, remains at rest, or moves uniformly in a straight line.

"This, the Inertia of matter, is proportional to the quantity of matter in the body. And it follows that some cause is requisite to disturb a body's uniformity of motion, or to change its direction from the natural rectilinear path."

The question of the cause of inertia has perplexed scientists for many centuries. Much work has been done to determine the cause, but to this day no specific cause has yet been generally accepted by the scientific community.

This article suggests two separate causes of inertia.

The first is based upon the proposition of the existence of aether, a sub-atomic substance that permeates the universe. This approach assumes that aether is a material substance that is subject to Newton's Laws of Motion. Thus, when aether interacts with material objects, it restrains their movement.

The second proposal contemplates that vibrations in the structure of atoms restrain the movement of atoms and atomic matter.

These two approaches fundamentally differ. One involves the interaction of aether with atomic matter, whereas the other is an intrinsic property of matter, namely vibrations.

Amitabha Ghosh poses this question: [2]

"...is the inertia of an object an intrinsic property of matter (irrespective of the presence of other matter in the universe) or is it nothing but the manifestation of the interaction of the moving object with the other matter present in the rest of the universe?"

Which concept is preferable? Or, might the cause of inertia be a combination of them both?

2. The Aether Approach

The aether approach is based upon the propositions that aether is a material substance, that it pervades the universe, that it is sub-atomic in size, and is subject to Newton's Laws of Motion.

The aether cause-of-inertia proposal is essentially derived from the author's cause-of-gravity concept that is set out in two articles: The Cause of Gravity: A Concept; [3] and Flowing Aether: A Concept. [4]

The cause-of-gravity concept is briefly as follows: Cosmic bodies expel aether cells into space where they condense into groups of aether cells that flow back into cosmic bodies. The expulsion and inflow are cyclic and continuous. The inflow is caused by a pressure difference between aether in cosmic bodies (lower pressure) and aether in space (higher pressure). The lower pressure in cosmic bodies is caused by the expulsion of aether cells. The process is like that of a household vacuum cleaner. It expels air and this reduces the air pressure in the machine below the ambient air pressure, thus causing air to flow into the machine. The one-way force of gravity is the result of the emitted aether cells being extremely small, tending to pass on through atomic matter, whereas the returning groups of aether cells are larger, tending to collide with atomic matter and push it toward cosmic bodies.

Both the gravity and the cause-of-inertia concepts are built upon the proposition that aether interacts with atomic matter in accord with Newton's Laws of Motion. Based on this assumption it is argued that aether that surrounds and interacts with atomic matter restrains its movement. Put another way, the presence of surrounding aether tends to keep matter in place, and force is required to overcome this restraint.

In effect, this proposal is akin to Mach's Principle which ties all movement in the universe to everything that exists in the universe. As stated by Ghosh:[5]

"'Mach's Principle' ...proposes that the inertia of an object to acceleration is due to the resistance generated by its interaction with the matter present in the rest of the universe."

However, for all practical purposes it is not necessary to look to the whole universe to comprehend the essence of the present cause-of-inertia proposal. The contribution of far-away aether to close-at-hand inertia must be miniscule. It is the direct interaction of atomic matter with adjacent aether that the present concept of inertia is aimed at.

In summary, the aether proposal is that aether is a material substance and, like any other material substance, is subject to Newton's Laws of Motion. Applying Newton's laws to aether offers an explanation for aether having a restraining effect upon atomic matter.

It is noted that two recent published articles advocate the interaction of aether and atomic matter as the cause of inertia. [6,7]

3. The Vibrations Approach

It is an undisputed fact that atoms vibrate. The proposal here is that the vibrations of atoms cause atoms and atomic matter to resist movement.

It is well known that rotating or spinning objects, such as flywheels and gyros, resist being moved off their planes of rotation. This observation is equally applicable to pendulums that oscillate back and forth. A good example is the Foucault pendulum. It maintains its plane of oscillation while the Earth rotates beneath it. It stands to reason that vibrating objects must have the same property. As stated by Eric Laithwaite: [8] " . . . an oscillating mass has all the properties of a wheel and mathematically may be so represented." In all these situations, there are back and forth oscillations of matter, each having its own plane of oscillation. This common factor ties planes of oscillation to the observed resistance to movement.

The centre of a body's inertia depends upon the movements and the masses of each individual part of the body. As stated by Thomson and Tait: [9, 10]

"The kinetic energy of any system is equal to the sum of the kinetic energies of a mass equal to the sum of the masses of the system, moving with a velocity equal to that of its centre of inertia, and of the motions of the separate parts relatively to the centre of inertia."

And:

the products of the masses of all the particles each into the square of its distance from the axis."

The argument here is that when a force moves an object, the movement changes the planes of vibration of virtually all the atoms that comprise the object. It stands to reason that the collective vibrations of the object's atoms provide the object with a measure of resistance to movement. This resistance is innate to atomic matter.

The above-cited Thomson and Tait treatise deals extensively with the physics of gyrostats. [13] The treatise examines the movements of frameworks that have gyrostats attached to them. The results show that the gyrostats have inertial effects on the frameworks. For those interested in further exploring the inertia proposal, you will find that the treatise is a treasure trove of information.

The inertia proposal is restricted to vibrations. Many scientists view atoms as containing spin, rotation and/or vortices. These phenomena involve reciprocal movements that provide resistance to changes of their planes of oscillations, just like vibrations. In view of this, why not extend this proposal to spin, rotation and vortices?

The answer lies in the necessity of atoms having structure. While vibrations are part and parcel of solid structures, spin, rotation and vortices are the antithesis of structure. Visualize matter in its solid state. In order to form solid structures, atoms must be firmly attached to each other. The idea of atoms being made up of spinning or rotating or swirling parts, runs counter to the property of stability of structures. In regard to atoms in their gaseous and liquid states, they need structure. The liquid and gaseous state atoms are in constant motion and collide with each other. They need structure to withstand these collisions. In addition, no matter what state atoms are in -- solid, fluid or gaseous -- they need the stability of structure to enable them to contain their immense energy. This is not to say that atoms cannot be structured so as to enable them to hold particles that spin, rotate or form vortices. However, in order to accommodate and contain these phenomena, elaborate structures would be needed. On the present state of knowledge of atoms, that prospect does not appear likely. Accordingly, the cause-of-inertia proposal is restricted to vibrations.

4. Conclusion

This article presents two separate versions of inertia. Each suggests a means of resistance that a force must overcome to cause movement. The author has not, as yet, found a convincing rationale that falsifies and therefore eliminates either of the two concepts.

It is suggested that the cause of inertia may be a combination of interaction of matter with surrounding aether and the innate property of matter, namely vibrations.

Might one be preferable as being more powerful than the other? Assuming that vibrations in atoms are an aspect of electromagnetism and a means of storage of the immense energy that atoms are known to possess, and that the pushing force of aether is likely small compared to these electromagnetic forces, it is suggested that vibrations in atoms are probably the dominant cause of inertia.

References

[1] W. Thomson (Lord Kelvin) and P.G. Tait, Principles of Mechanics and Dynamics, Part One, (Constable and Company, Limited, London, 1962), pp. 222-223, para. 216.

[2] A. Ghosh, Mach's Principle and the Origin of Inertia, edited by M Sachs and A.R. Roy, (C. Roy Keys, Inc., Montreal, 2003)), p. 13.

[3] D.W. Shaw, Phys. Essays, 25, 66 (2012).
[4] D.W. Shaw, Phys. Essays,, 26, 523 (2013).
[5] Ref. 2.
[6] G. Boersma, Phys. Essays 27, 259 (2014).
[7] M. Sato and H. Sato, Phys. Essays, 28, 95 (2015).
[8] E. Laithwaite, An Inventor in the Garden of Eden, (Cambridge University Press, Cambridge, UK, 1994), p. 243.
[9] Ref. 1, at p. 259, para. 280.
[10] Ref. 1, at pp. 259-260, para. 281.
[11] Ref. 1, chapter 2, pp. 219-439, paras. 205-368.

On the Structure of Atoms

Duncan W. Shaw
1517 Angus Drive, Vancouver, B.C., Canada, V6J 4H2.
e-mail: duncanshaw@shaw.ca

The idea of atoms being structured by particles that are physically attached to each other, like connected pieces of a Meccano set, has been around a long time. As a broad proposition, it has a certain appeal. However, it is not in vogue at the present time, perhaps because it is difficult to envision its precise details, such as how might the particles be shaped and how can they physically form structures. Present day science conceives that forces, such as the strong force and electromagnetic forces, provide atoms with structure. What if the force concept turns out to be wrong? Then, the physical attachment approach comes back into focus.

This article brings the physical attachment concept back into consideration. It suggests that atoms are structured by means of their constituent particles being physically connected to each other. Implications of the concept are discussed in the article.

1. Introduction

This article is submitted to the John Chappell Natural Philosophy Society for discussion purposes at the society's 2015 conference.

The article relates to the structure of atoms. It proposes a concept of atomic structure based upon the underlying premise that the particles that constitute atoms are physically joined to each other.

2. The Concept

The proposed concept is that the atom is a mechanical structure in the sense that all its particles are physically connected together. This concept encompasses the particles in the nucleus and the particles that occupy the volume that extends from the nucleus to the atom's outer perimeter. The particles physically anchor each other such that when combined together they form a solid structure, that structure being the atom.

This proposition has many implications. One is that if the concept is correct, it may eliminate the present generally accepted concept that the "strong force" is the means by which the constituent parts of the nucleus are held together. Another is that it may eliminate the concept that "electromagnetic forces" bind the constituent parts of the atom together. Why? Because actual physical connections provide the cohesion, rather than forces.

3. The Need for Structure

For the following reasons, it is suggested that atoms need to have solid structure. The word "solid" is used in the sense of particles being firmly but flexibly held together so as to form a structure. The word "structure" is used in the mechanical sense – as in in regard to a building or a bridge – where a stable structure is formed by its constituent parts being physically joined together.

Why must atoms have solid structure? To answer this question, it helps to visualize the four states of matter, namely solids, liquids, gasses and plasmas. To start with, consider atoms in their solid state, that is, atoms that are the constituents of solid structures such as buildings and bridges. Without solidly structured atoms, it is fair to say that buildings and bridges would collapse. Indeed, they could not even be constructed. Simply put, if atoms do not have solid structure, they cannot form solids with other atoms.

4. Implications

There are numerous potential implications to the mechanical structure proposition. Here are several.

The strong force. Present day physics considers that the nucleus is held together by what is called the "strong force". Does this force in fact exist? There is good reason to conclude that it does not. Keep in mind that the concept of the strong force is based upon the fact that it takes considerable force to break the grip of whatever holds the nucleus together. It is a matter of inference that there must be a strong force that holds the particles of the nucleus together. But, if one accepts the proposition that the particles are physically connected and thereby resist being broken apart, then the inference of a strong force that does the binding is not needed.

Electromagnetic forces. In addition to the strong force, electromagnetic forces are invoked to explain the internal cohesion of the atom, i.e., why the atom does not fly apart. As with the strong force, the concept of physically linked particles may eliminate or perhaps supplement the explanation that electromagnetic forces provide binding structure.

Orbit and cloud concepts. There is a fundamental problem with the "solar system" and the "cloud" concepts of electrons in atoms. The problem is that these concepts connote a lack of structure. How can electrons that are in orbit or which form a cloud physically latch on to the clouds or orbiting electrons of neighboring atoms? How can they do so and form solids? It simply does not make sense. It stands to reason that atoms must physically attach to each other to form solids. If they cannot firmly attach, they will slide on by each other and not form a solid. The orbital and cloud concepts are incompatible with the proposed structure concept.

Storage of energy. There are various methods said to be the means of storing energy in atoms. The main contenders are spin, rotation, vortices and vibrations. Which is preferable? It is suggested that vibrations is the simplest and the most likely candidate. Spin, rotation and vortices are the antithesis of structure. Think of constructing a house with materials that spin, rotate or swirl in vortices. Not easy to imagine. Parts that spin, rotate or swirl might be attached to buildings, but they do not form the essential structures that provide buildings with their strength. On the other hand, vibrations are part and parcel of all structures. There is no building that does not have vibrations, just as there are no atoms that do not have vibrations. It is suggested that Occam 's razor argues for the simplest method of storing energy, that being by way of vibrations.

Shape of particles. The physical linkage concept raises the problem of the actual make-up of particles that connect with each other. This is probably the most fundamental issue associated with the physical connections concept. Some physicists theorize that particles called "gluons" do the job, and perhaps they do. But, on the assumption that particles in the nature of gluons or some fundamental particles of similar properties in fact exist, there remains the question of how these particles might accomplish mechanical linkage. It is suggested that to effect physical linkage, particles must be shaped such that they can hold onto each other, much like pieces of a Meccano set. For discussion purposes, consider the possibility of such particles being shaped like the letter "C" or the letter "S". Might particles of these configurations combine by random interactions and set in train the formation of all the substances that make up the universe? Has any one got other suggestions of the shape of fundamental particles that might be capable of physical attachment? Ideas bearing on this matter would be most welcome.

Shells. Visualize an atom with its constituent particles being physically held in place. Visualize these particles fitting together as structures. Consider the structures being like shells, with each shell having set numbers of particles that fit together and occupy set positions. Consider the particles vibrating and, by successive collisions, transferring their vibrations to neighboring particles and, in doing so, circulating the energy of the vibrations continuously inside the shells. Might this be the essential means of storage of the immense energy of atoms? See the Storage of energy section above.

Impression of orbiting. Energy that is circulating in electron shells might give the false appearance of being electrons in orbit. Assuming that electrons stay put in their respective shells, might their vibrations, as distinct from the electrons themselves, circulate in the electron shells, thus creating the incorrect impression of electrons being in orbit?

Electrons framework. The volume of the part of the atom that extends from the nucleus to the atom's outer perimeter is far greater than the collective volume of the electrons that are situate in that space. This fact raises the question of how the electrons can be held in structured positions and be physically connected to each other and to the nucleus. A suggested response to this question is that a subatomic substance occupies the volume from the nucleus to the outer perimeter that provides structural framework -- in the form of shells -- that holds electrons in their positions and provides the necessary attachments. Might the subatomic substance be aether?

Expulsion and absorption of electrons. Vibrational energy of atoms can become so elevated that the vibrations force electrons from their structured positions. When this happens, electrons may be expelled from their shells or caused to change positions from one shell to another. Each expulsion or position change would open up gaps in electron shells and, when this occurs, the pressure of surrounding particles may force electrons that are in the vicinity to fill the gaps.

The outer reaches of the atom. The physical attachment approach to the atom's structure has implications in regard to the borders of the atom. It is suggested that atoms must have boundaries that permit them to physically connect to neighboring atoms. Without firm attachments between atoms, it seems evident that the solid state of matter cannot be formed. As for the liquid state of matter, there must at least be partial connections that permit the reduced level of cohesion that is characteristic of fluids. All atoms, including gasses, must at least have outer surfaces of substance that accommodate collisions and rebounding. What, then, might constitute the outer borders? Consider this possibility: that the atom's outermost shell forms the atom's physical perimeter. If the assumption that aether forms shells is correct (see the Electrons framework section above), it appears reasonable that shells would have the capacity to physically connect with the shells of neighboring atoms. Out of interest, might this idea provide an explanation for the phenomenon of surface tension?

Conclusion

Physical attachment of the particles that make up the atom is seen as a possible replacement to the present-day concept that forces are the basis of the atom's cohesion. The proposal that atoms must have mechanical structure is based upon propositions that: (1) without atoms being solidly structured, there can be no solid structures such as buildings and bridges; (2) solid structure is necessary for atoms to contain, without breaking apart, the immense energy they possess; and (3) strength is required for atoms to withstand collisions between atoms. The physical attachment concept of the structure of atoms has far-reaching implications, several of which are raised in this article.

Making Waves

Scott Wall
8 Harrop Ave, Georgetown, Ont, Canada
e-mail: scottwall@alumni.waterloo.ca

The consensus of oceanographers is that ocean waves are predominately caused by wind and supplemented by a few other sources. The following are said to be the causative forces behind ocean waves:

- wind
- gravitational attraction of the sun and moon
- wave shoaling (by entering shallower waters)
- wave refractions (from bathymetry and currents)
- variations in wind and atmospheric pressure (seiches)
- displacement waves (from boat wakes, landslides, etc.)
- interactions with sea ice and icebergs (dampening effect)
- gravity (dampening effect)
- viscosity (dampening effect)

This paper intends to examine the efficacy of some of these forces in generating the gamut of ocean waves.

1. Introduction

Wind and gravity are the bedrock of the ocean wave theories for oceanographers. Due to the uncertainties of the subject matter, Blair Kinsman lamented the situation as follows: "So long as you don't do something absolutely absurd, you are bound to get "oceanographic level" agreement. Some day somebody is going to take a close, accurate look, and the whole "agreement" will go sky high." [1] Some day somebody will consider the role of electromagnetism on the ocean.

2. Wave Coherency

According to Dr. Gerald Pollack, the surface of water has an electrostatic crystalline structure which he calls an Exclusion Zone mosaic [2, chapter 16]; he refers to this as an elastic sheet model. This structure allows water bugs and even some lizards to walk on water. Pollack envisions that this sheet model explains why tsunami waves can swiftly circumnavigate the earth several times before dissipating. Pollack also implies that, in general, ocean waves follow the elastic sheet model. From my observations, I would suggest that this model extends to all bodies of water, not just oceans. Disturbances on water surfaces appear to propagate much like a jiggling bowl of Jello; changes to the area of the surface seem to be damped by a resistive force, hence, maintaining the wave structure of the water.

3. Scalability of Wave Formation

Oceanographers and mathematicians are only interested in water waves on oceans, seas and "Great" lakes. They have no interest in waves of lesser bodies of water. From an economics point of view, ocean waves can have a significant impact on residential and commercial areas, particularly commercial shipping. Occasionally, they will use wave generators to mimic what they suspect is happening, but in doing so, they have are forcing the causative agent.

If you consider a body of water to be the laboratory, large bodies of water have far too many unknowns and are simply not suitable for studying the fundamentals of wave formation. Reducing the size of the laboratory is quintessential to gleaning the underlying processes. Hence, the smaller is the body of water, the fewer will be the unknowns.

Let's begin our investigation by getting extreme and starting with a bowl of water. There are four ways to disturb the water's surface:

1) Move the bowl, e.g., landslide into water
2) Blow on the bowl, e.g., wind
3) Drop something into the bowl, e.g., displacement wakes
4) Place a statically charged balloon directly above the water.

The first three need no explanation, but, for the fourth, we will again refer to Pollack as he explains evaporation: "This lifting force may be electrostatic. You may recollect that vesicles carry a net negative charge. Negative charge alone cannot explain lift; however the earth bears negative charge as well. The earth's negative charge may repel the vesicles, pushing them upward." [2, Chapter 15]

The voltage gradient of the atmosphere is said to usually measure between 100 volts / meter in the summer up to 300 volts / meter in the winter [3]. Does this voltage gradient provide some lifting force? Charles Lucas Jr. has taken the four fundamental laws of physics and derived a force equation for the core fundamental electrodynamic force. [4] His paper conclusively shows that gravity is a subset of electrodynamic forces.

4. Tide

Everyone has heard that tides are caused by the gravitational attraction of the sun and moon. I've never heard anyone comment on the absurdity of this statement. How on earth can the gravity of the minuscule moon out-muscle the gravity of the earth and substantially lift huge quantities of water from the oceans? Similarly, why would the distant sun have any gravitational effect on the oceans? To illustrate my point, given the gravitational force equation of $F = GmM / R^2$, the ratio of the moon's gravitational force versus the earth's gravitational force at the surface of the ocean works out to $M_e * R_m{}^2 / M_m * R_e{}^2$. Given the following constants:

M: moon = 7.35 e22 kg; earth = 5.88 e24 kg; sun = 1.99 e30 kg

R (distance to ocean surface): moon = 384403 km; earth = 6378 km; sun = 149597870.7 km

Calculating these ratios yields the following: the attraction of the earth is 290,391 times greater than that of the moon and 1,624 times greater than the force from the sun. At the ocean's surface, the gravitational effect from both of these bodies should be negligible. However, the electrical attraction is a different matter all together. Proponents of the Electric Universe point to electrical connections between the moon and the earth, in addition to the sun and the earth. The sun has a positive charge; the earth, a negative charge. Since structured water also has a negative charge, it is pushed away from the earth and drawn towards the positive charge of the sun. This is very much like the bowl of water being drawn towards the balloon.

An electrical force could also explain the antipodal tide, i.e., the high tide on the side of the earth furthest from the moon. Diagrams of the earth's magnetosphere correspond to the crude schematics of the tidal forces.

5. Barometric Pressure

Meteorologists tell us that barometric pressure is the weight of the atmosphere and fluctuations are due to cold or warm masses of air. Immanuel Velikovsky has raised some objections to this view: "... there are daily variations in the height of the barometer, culminating in two maxima and two minima during the course of 24 hours. The heating of the sun can explain neither the time when the maxim appear (10 am and pm) nor the time of the minima (4 am and pm) of these semidiurnal variations." [5]; "The lowest pressure is near the equator, in the belt of the doldrums. Yet the troposphere is highest at the equator, being on the average about 18 km high there; it is lower in the moderate latitudes, and only 6 km. high above ground at the poles." [5] Since the fundamental theory of this paper is "Gravitation is an electromagnetic phenomenon", I take this to be supportive of the view that barometric pressure is also an electromagnetic phenomenon as it is merely the "gravitational weight" of the atmosphere with gravity also being an electromagnetic phenomenon [4].

6. Storm Surge

Storm surge is water that is pushed onto the shore by hurricanes. There are three mechanisms that contribute to surge [6]:

1) action of winds piling up water (~ 85 %)

2) waves pushing water inland faster than it can drain off (5 - 10%)

3) low barometric pressure sucking water higher into the air (5 - 10%).

Since barometric pressure is an electromagnetic phenomenon as deduced in section 4 of this paper, electromagnetic forces are responsible for at least 5 to 10 percent of the water being sucked higher in a storm surge.

It is highly possible that a portion of the "action of winds piling up water" may also be attributed to waters being lifted by barometric influences; this would increase the electromagnetic contribution to the surge.

7. Dust Devils and Water Spouts

Water spouts and dust devils have the appearance of being miniature tornadoes, i.e., intense columnar vortices, with water spouts occurring over water and dust devils being land-based.

The electrical nature of dust devils has been confirmed and has been measured to be in excess of 4,000 volts per meter. [7] This article quotes a NASA report: "Dust devil winds carry the small, negatively charged particles high into the air; while the heavier, positively charged particles remain near the base of the dust devil."

Water spouts are quite common and are deemed to be caused by vertical wind shear [8]. It seems unlikely that a vertical wind shear could form a tight funnel while also lifting water from the sea up a considerable distance to the clouds and counteracting the intense centrifugal forces. What force constrains the water to the spout? Considering that water spout's are similar to dust devils, it's more likely that electric charge is at the heart of the water spout.

Sometimes the spout carries the water from the sea up to the parent cloud [8]. Structured surface water has a negative charge, as does the earth's surface [2]. Perhaps the voltage gradient of the atmosphere would be enough to lift the water up to the cloud.

8. Tornadoes

Wal Thornhill describes tornadoes as a slow discharge mechanism from cloud to ground. "Measurements of the magnetic field and earth current near touchdown of a tornado shows that it is electrically equivalent to several hundred storm cells." [7] Edward Lewis has been studying plasmoids since 1990 and states definitively that "tornadoes are a plasmoid phenomena, which are basically an electrical-magnetic phenomena." [9] In this article, Lewis also discusses the experiences of several people caught in the eye of a tornado.

These encounters are interesting, but I have to say that the most fascinating story is that of Mark Spann. In 2011, Spann was sitting in his bathtub with his roommate and neighbour with the bath curtain covering them when the tornado's eye passed directly overhead. Here is how he described the experience: "two distinctly sharp pressure gradient changes as the inner plasma sheath passed over us" [10] and "The sensation of weightlessness began just a few seconds before the center vortex and highest winds reached us. Start-to-finish the weightlessness lasted less than ten seconds, but during it I got the same queasy 'butterfly's-in-the-stomach' feeling that I would get riding the fast roller coaster at Six Flags over Georgia as a kid." [10] Furthermore, "Adrenaline flow aside, one would not experience such a sensation had the 'weightlessness' phenomena been due to the

Bernoulli effect, wind pressure gradient anomalies, vortex suction, etc., and of course, had some such a wind-pressure laminar flow sheath been responsible sufficient to lift a human body out of a sink-hole, without a doubt the shower curtain would have been ripped out of our hands and vanished up into the funnel when the adjacent exterior wall blew out and the roof vanished up into the funnel, leaving us exposed to 200+mph winds on the back-side of the vortex as it passed on its way away from us to the east." [10]

I suspect that the weightlessness was a result of electromagnetic lift being provided to the internal organs responsible for the sense of weightlessness. The responsible organs, be it the inner ear, stomach or brain, are filled with fluid which consists of negatively charged structured water. Electromagnetic forces lifted the organs. The curtain did not experience this force.

9. Unusual Precipitation

Since ancient times, there have been stories of many unusual creatures falling from the sky, such as snakes, frogs, fish, etc. One explanation provided was that the creatures "had been lifted from a nearby ocean or lake by waterspouts or tornadoes, carried overland and dumped to the ground." [11]

This explanation doesn't even bother trying to explain how waterspouts can lift frogs and snakes but not dirt or mud; the deluges have a high density of creatures. As with Mark Spann and his tornado, living creatures contain structured water which would experience a lifting effect from the negative charge in the waterspout.

10. Rogue (Freak) Waves

Rogue waves or Freak Waves are enormous waves, often exceeding 30 meters in height, which sporadically occur in all oceans. They are largest where opposing currents meet, but this is not a precondition of their formation. Mainstream oceanographers offer the following explanations [12]:

1) Ocean wave theory does not permit waves with these heights; therefore, the mariners reporting these waves are exaggerating their height.

2) The waves are exhibiting behaviour similar to quantum mechanics; the rogues are "stealing energy from their neighbours".

Enough irrefutable evidence of rogue waves has been gathered, that the first explanation has been completely disproven. The second explanation ventures into meta-physics, invoking unknown, mysterious forces and does not relate well to the physical world in which we live. Perhaps, structured water and electromagnetism could help explain the unexplainable.

Consider that both currents consist of highly viscous structured water that is maintaining a relatively constant surface area. If each of these currents is severely disturbed by various forces, then the cumulative superpositioning of the waves could become quite large. One of the disturbing forces may be an anomalous electromagnetic pulse. Mostly, rogue waves appear when electrical storms are present, adding to the plausibility of an electromagnetic influence.

11. Conclusion

The structural "Exclusion Zone" [2] of water provides the cohesion for the waves. The negative charge of this water when combined with the atmospheric voltage gradient creates an electromagnetic force that is a major contributor to ocean waves, especially tides and water spouts. Together, structured water and electromagnetism play a major role in oceanography and related phenomenon.

References

[1] Kinsman, Blair, 2012, "Wind Waves: Their Generation and Propagation on the Ocean Surface"

[2] Pollack, Gerald, 2013, "The Fourth Phase of Water: Beyond Solid, Liquid, and Vapour"

[3] Meridian International Research, 2005, http://www.meridian-int-res.com/Energy/Atmospheric.htm

[4] Lucas Jr., Charles W., 2011, "The Universal Electrodynamic Force", Natural Philosophy Alliance 18th Annual Conference, College Park, MD.

[5] Velikovsky, Immanuel, 1946, "Cosmos without Gravitation: Attraction, Repulsion and Electromagnetic Circumduction in the Solar System; Synopsis".

[6] Masters, Jeffrey, 2015, Weather Underground website, http://www.wunderground.com/hurricane/surge.asp

[7] Thornhill, Wal, 2004, "Electric Dust Devils", http://www.holoscience.com/wp/electric-dust-devils/

[8] Smita, 2011, Marine Insight website, http://www.marineinsight.com/marine/8-facts-about-water-spouts-at-sea/

[9] Lewis, Edward, 2013, "Tornadoes, Plasmoids and Ball Lightning Identification Evidence", http://www..thunderbolts.info/wp/2013/02/01/tornadoes-plasmoids-and-ball-lightning-identification-evidence/

[10] Spann, Mark, 2015, private communication

[11] Dennis, Jerry, Wolff, Glenn, "It's Raining Frogs and Fishes: Four Seasons of Natural Phenomena and Oddities of the Sky"

[12] National Geographic, 2004, Rogue Waves